ROBERT M. KOERNER AND DAVID E. DANIEL

FINAL COVERS FOR SOLID WASTE LANDFILLS AND ABANDONED DUMPS

Published by
American Society of Civil Engineers
1801 Alexander Bell Drive
Reston, Virginia 20191-4400

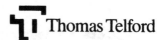

Co-published in the UK by
Thomas Telford Publications
Thomas Telford Services Lt
1 Heron Quay
London E14 4JD, UK

Abstract:

This book presents the essential elements for the design of final covers for solid waste landfills and abandoned dumps. Chapter 1 gives an overview and presents selected aspects of regulations in the US and Germany; countries where the regulations are the most advanced. Chapter 2 gives detailed information on each of the six individual layers of final cover systems. Chapter 3 contains example cross sections of nine candidate final cover designs. Chapter 4 presents a rational step-by-step procedure of a water balance analysis. Chapter 5 contains analytic details on cover soil slope stability analyses, with worked-out examples. Chapter 6 presents related final cover designs and emerging concepts. Chapter 7 includes additional considerations (quality control, lifetime of materials, post-closure details, etc.) and a summary of the book.

Library of Congress Cataloging-in-Publication Data

Koerner, Robert M., 1933-
Final covers for solid waste landfills and abandoned dumps / Robert M. Koerner and David E. Daniel.
p.   cm.
ISBN 0-7844-0261-2
1. Landfill final covers. I. Daniel, David E. (David Edwin), 1949-   . II. Title.
TD795.7.K64 1997   97-20893
928.4'4564--DC21   CIP

Any statements expressed in these materials are those of the individual authors and do not necessarily represent the views of ASCE, which takes no responsibility for any statement made herein. No reference made in this publication to any specific method, product, process or service constitutes or implies an endorsement, recommendation, or warranty thereof by ASCE. The materials are for general information only and do not represent a standard of ASCE, nor are they intended as a reference in purchase specifications, contracts, regulations, statutes, or any other legal document.
ASCE makes no representation or warranty of any kind, whether express or implied, concerning the accuracy, completeness, suitability, or utility of any information, apparatus, product, or process discussed in this publication, and assumes no liability therefore. This information should not be used without first securing competent advice with respect to its suitability for any general or specific application. Anyone utilizing this information assumes all liability arising from such use, including but not limited to infringement of any patent or patents.

Photocopies. Authorization to photocopy material for internal or personal use under circumstances not falling within the fair use provisions of the Copyright Act is granted by ASCE to libraries and other users registered with the Copyright Clearance Center (CCC) Transactional Reporting Service. Requests for special permission or bulk copying should be addressed to Permissions & Copyright Dept., ASCE.

CONTENTS

Chapter 1.   Introduction

Chapter 3.   Final Cover System Cross Sections

Chapter 4.   Water Balance Analysis

Chapter 7.   Other Considerations and Summary

PREFACE

The number of solid waste landfills and abandoned dumps worldwide is staggering by anyone's count. While the situation is most acute in industrialized nations, developing nations are also involved since they have been dumping grounds on many occasions. Clearly, solid waste should be properly treated and then disposed of in an environmentally safe and acceptable manner. Yet, such treatment dwarfs the economic capability (and/or willingness) of even the most financially healthy of nations. An admittedly poor second choice to waste treatment, is waste containment. Furthermore, the key to containment is often a final cover over the landfill or dump.

We say final cover because all indications are that such covers will be in place indefinitely. One would be naive to think that the world economy would suddenly have the trillions of dollars necessary to treat and remediate all of the known (and many as yet unknown) sites. Hence, the orientation of this book toward *final covers for solid waste landfills and abandoned dumps.*

An overview of regulations in the United States and Germany is covered in the first chapter, but regulations certainly do not dominate the designs set forth in the remaining six chapters. In what is hoped to be a performance-based design approach, we have set forth successive chapters on the following aspects of final covers:

- Individual components of candidate cover systems. Detail is presented on the surface, protection, drainage, barrier, gas collection, and foundation layers for the entire range of natural soil materials and geosynthetics.
- Example cross sections of final covers. Included in this chapter are nine candidate cross-sections of final covers. There are three each for hazardous waste, nonhazardous waste, and abandoned dumps. Most are illustrated for both arid and humid sites.
- Details of a water-balance methodology. A technique based on first principles is developed and illustrated by a worked-out hand calculation. It is then compared and contrasted to the HELP computer model.
- Theory and design examples on slope stability. Slope stability is analyzed for uniform and tapered cover soil thicknesses, equipment loads, seepage, seismicity, and veneer reinforcement. Worked out examples illustrate each segment.
- Elements of other designs and emerging systems. Capillary barriers, leachate recycling, and mine residuals are described along with emerging materials. In this latter group are erosion control

materials, geofoam, shredded tires, and alternative liners. Selected case histories of field performance of final covers are also included.
- Related considerations and summary. Included is discussion on quality control and quality assurance, lifetime of natural soils and geosynthetics, warrants, post-closure issues, and a summary.

The overlying tone of the book is for the design to be performance driven for the site specific situation and not to be prescriptive. While prescriptive details are understandable from a regulatory perspective, they are not the approach which typically generates an ideal site-specific design. When considering a site-specific final cover, the designer must focus on the benefit of a particular cross section vis-a-vis its cost. This is the essence of design, i.e., to optimize the benefit/cost ratio for the site-specific decision. In this book, many options are given to the reader. It would be presumptuous for us to say that all options are covered herein, yet, we feel that the main topics are treated in an open and unbiased manner. We wish you the best in your present and future work regarding the design, installation, and performance of solid waste landfills and abandoned dump final covers.

Robert M. Koerner
David E. Daniel

ACKNOWLEDGMENTS

While this effort was not federally supported, we would be remiss if we did not express sincere appreciation to the U.S. Environmental Protection Agency (EPA) for past support. Past project officers Robert E. Landreth and David A. Carson have been extremely helpful in this regard. Federal agency projects from the EPA, the Department of Energy, and the Department of Defense (e.g., U.S. Army Corps of Engineers) have helped both of us shape our opinions to the point where we felt this book was necessary and could be significant to the geoenvironmental community at large.

Sincere appreciation is also extended to Dr. Te-Yang Soong and Ms. Marilyn Ashley, both of the Geosynthetic Research Institute, for their interaction and cooperation.

CHAPTER 1

INTRODUCTION

This book on final covers refers to engineered cover (also called "cap") systems that are placed over solid waste landfills, abandoned dumps, or contaminated materials. All types of solid waste materials (nonhazardous and hazardous) are considered. In the most general context, Figure 1.1 depicts the various applications.

Covers placed over landfills, as in Figure 1.1a, are typically multi-component cover systems that are constructed directly on top of the waste shortly after a specific unit or cell has been filled to capacity. The waste may be low-level radioactive waste, hazardous waste, nonhazardous industrial waste, municipal solid waste, incineration ash, construction/demolition waste or some other type of waste. Modern waste management units, so-called "landfills," are almost always constructed with a bottom liner system that includes a leachate collection layer. Leachate is the contaminated liquid that drains from the waste material. The final cover system is intended to (a) control infiltration of water into the landfill thus minimizing leachate, (b) control the release of gases from the landfill, and (c) provide for a physical separation between the waste and environment for protection of public health. As will be seen later, the rules and regulations for landfill covers vary considerably in different countries and from one type of waste to another, but the engineering principles are essentially the same.

Cover systems placed over abandoned dumps, as in Figure 1.1b, are similar to those constructed over newly completed landfills. They are usually multi-component systems that are designed to fulfill the same functions mentioned earlier for covers placed over a recently completed landfill. However, some of the design challenges are different because the type and condition of the underlying waste is usually not very well defined. For instance, the thickness, lateral extent and composition of the waste are often poorly defined and the differential settlement of the final cover can be extreme if the buried waste contains large voids. Furthermore, the typical absence of a liner system beneath abandoned dumps makes the proper

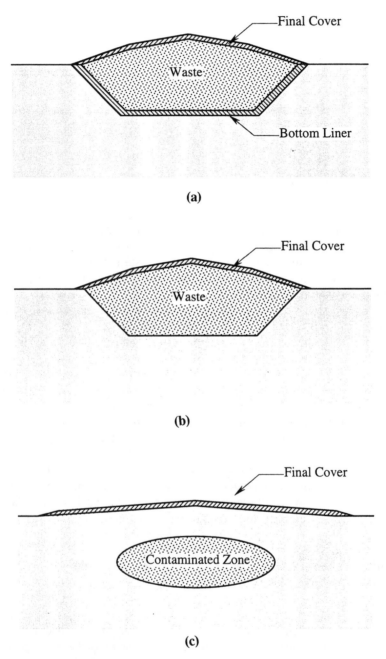

FIG. 1.1.   Final Covers on Different Types of Waste Described in this Book

functioning of the cover absolutely essential. Capping may be undertaken as the sole control measure for an abandoned dump, but in most cases the capping is part of a broader program of corrective action that may include installation of leachate recovery wells or gas collection wells, as well as a lateral containment system such as vertical cutoff walls. Figure 1.2 illustrates how a final cover system might be used in conjunction with collection wells and a vertical cutoff wall.

Cover systems are often placed at locations where contaminated materials are found at the ground surface or at shallow depths. Sources of the contamination include leaking underground storage tanks, leaking pipelines, dewatered ponds or lagoons, surface spills, and a variety of other environmental situations. Corrective action at contaminated sites often includes a variety of activities, including installation of groundwater recovery wells, installation of soil vapor extraction systems, and installation of physical barriers, such as vertical cutoff walls, that are designed to contain the waste and its leachate. In virtually all cases, the principal objective of the cover system at corrective action sites is to limit percolation of water into the subsurface, since such percolation would create additional contaminated liquid that would add to the scope of the corrective action program. In some cases, the cap may also serve as a physical separator between the contaminated materials and the surface environment (for instance, to prevent animals from ingesting contaminated soil). Cover systems for corrective action projects often are similar to those for landfills, since the principal functions of final cover systems are very similar for engineered landfills and

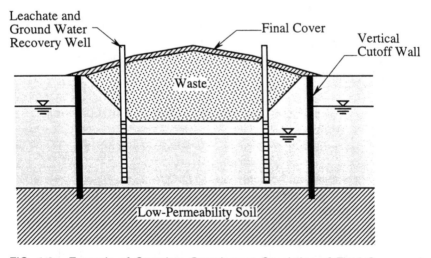

FIG. 1.2. *Example of Complete Containment Consisting of Final Cover and Vertical Cutoff Walls of Abandoned Dump Including Recovery Wells*

abandoned dumps. However, their areal coverage may be quite extensive as illustrated in Figure 1.1c.

The *primary* objective of most final cover systems over solid waste materials is to prevent the percolation of meteoric water into underlying waste or contaminated soil. Therefore, properly designed and constructed covers arguably become, after closure, the most important feature of a landfill. To function properly, the cover must be designed and constructed to provide long-term minimization of the movement of water from the surface into the closed waste mass beneath. Where the waste mass lies entirely above the zone of groundwater saturation, a properly designed and maintained cover can prevent, for all practical purposes, the entry of water into the underlying waste mass, and thus essentially eliminate the formation and migration of leachate. Final cover systems will never eliminate the need for liners and leachate collection systems in landfills, but if properly designed, constructed, and maintained, final cover systems can be the key component in safely managing liquids in landfills. A *secondary* objective of final cover systems is to prevent and capture air-polluting gases from escaping into the atmosphere. This issue has taken heightened importance in recent years.

As illustrated in Figure 1.3, the usual components, or layers, within a final cover system are the surface layer, protection layer, drainage layer, hydraulic/gas barrier layer, gas collection layer, and foundation layer. This book will revolve around the proper use and design of these individual layers vis-a-vis waste-specific and site-specific situations.

1.1 PURPOSE OF THE BOOK

The purpose of this book is to provide detailed guidance on the design of final cover systems for engineered landfills, abandoned waste dumps, and other types of correction action projects. The book is intended to be used by the following:

- Owners and/or operators of landfills and by principal responsible parties (PRPs) in corrective action projects to determine the minimum requirements for cover systems and to identify the major design elements.
- Engineering designers of final cover systems to establish the appropriate criteria and methods for design and management.
- Federal, state, and local regulators to confirm that the appropriate factors have been considered and that suitable methods of analysis and design have been employed.
- Interested members of the public, e.g., individuals living near landfills or corrective action projects, to provide an overview of

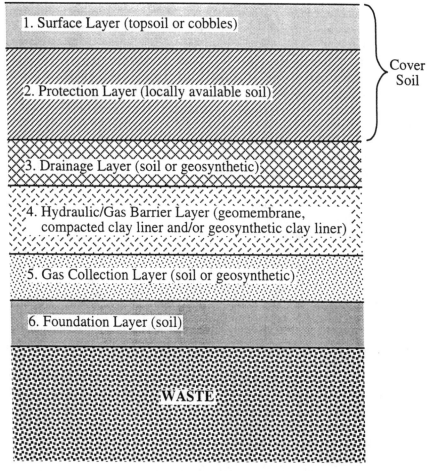

FIG. 1.3. *Six Typical Layers to Be Considered in Final Cover Design that Form the Focus of this Book*

the objectives of final covers and the methods by which those objectives are met.

- Construction quality control (CQC) and construction quality assurance (CQA) personnel to identify major design considerations and issues.
- Educators for the teaching of design and analysis of final cover systems to students in geoenvironmental engineering and related disciplines.

The remainder of Chapter 1 focuses on regulatory requirements for final covers, individual components of final covers, and other general issues related to covers. Chapter 2 provides a detailed description of the individual components of final covers. Chapter 3 presents detailed examples of cross sections of cover systems for various classes of solid waste materials. Methods for calculating rates of water percolation through final covers, a "water-balance analysis," are described in Chapter 4. Slope stability is a critical issue for final covers, and details for assessing slope stability are described in Chapter 5. Additional design and emerging concepts regarding final covers are described in Chapter 6. The very important considerations of construction quality control and quality assurance are discussed in Chapter 7. Also, such issues as performance monitoring, warrants, and escrowed funds are discussed in Chapter 7 along with summarizing comments.

Before beginning a detailed description of covers, however, it is important to note that engineered landfills, abandoned waste dumps, and waste-remediation projects are discussed in federal, state, and local regulations in many industrialized countries in the world. Indeed, to obtain a construction permit, the prevailing regulations must be followed. Yet, to the authors, the status of cover regulations for engineered landfills and waste dumps is not consistent nor (in some cases) logical. This inconsistency exists between countries and even within the same country. Regulations between agencies can be contradictory, leading to less than best-available-technology and/or excessive cost without commensurate benefits.

Even further, regulations rarely keep up to the state-of-the-art, let alone to the state-of-the-practice. New methods, new materials, new concepts, etc., should be incorporated when and where they are applicable. This is usually accommodated in regulations under the concept of "technical equivalency." However, the establishment of technical equivalency is often time-consuming and difficult to achieve, and is sometimes met with skepticism and/or distrust on the part of the regulator(s) involved in the decision.

For the above reasons, this book is purposely not regulatory driven. The book is based on the knowledge and experience of the authors and is designed to help achieve a safe and secure final cover facility in light of the benefit/cost of the particular system utilized. The selected cover design hopefully can be characterized as best available technology (BAT) in light of a particular site-specific situation.

Yet, to be oblivious to regulations is simply not realistic. Therefore, the next section presents elements of regulations in the United States and Germany. These two countries have arguably the most carefully considered and crafted regulatory documents in the world with respect to final cover applications. The U.S. Environmental Protection Agency (EPA) and the German Umweltbundesamt (UBA) have worked together on waste containment issues over the years. Yet, it is understandable that differences exist.

The proceedings of a workshop on similarities and differences between the United States and Germany for the geomembrane component of barrier systems is available (Corbet et al. 1997).

1.2 LINER AND COVER REGULATIONS IN THE UNITED STATES AND GERMANY

Regulations for liner systems **in the United States** (including closure and post-closure regulations) fall into two broad categories:

- New waste containment facilities (both hazardous and nonhazardous), which are regulated at the federal level under the Resource Conservation and Recovery Act (RCRA).
- Abandoned dumps and other contaminated sites that require corrective action and are regulated under the Comprehensive Environmental Response, Compensation, and Liability Act (CERCLA), commonly known as Superfund.

RCRA and CERCLA regulations are very different from one another. The RCRA regulations are generally much more specific and detailed, and in some cases offer little flexibility to the designer. CERCLA regulations, in contrast, offer little specificity and instead simply state general goals and objectives for final cover systems.

Under RCRA, hazardous and nonhazardous solid waste landfills are regulated differently. Hazardous waste landfills are regulated under "Subtitle C," and nonhazardous waste landfills are regulated under "Subtitle D." Furthermore, regulated nonhazardous solid wastes can be of two types: industrial waste and municipal solid waste (MSW). Municipal solid wastes are regulated under Subtitle D of RCRA; hence, MSW landfills are known as "RCRA Subtitle D" landfills. The regulation of nonhazardous industrial waste is ambiguous at the federal level and appears to fall somewhere between Subtitle C and Subtitle D wastes.

In the United States, landfills are regulated at the federal level via RCRA but regulation is also implemented at the state level. In most states, the state regulatory agency has the actual authority for implementing RCRA solid waste landfill permitting and compliance monitoring. Many states have their own set of regulations, which cannot be any less stringent than federal regulations. Thus federal regulations set what is known as "minimum technology guidance," or MTG. Some states, e.g., Texas, have made it their policy not to implement any landfill regulation that is more stringent than federal regulations. Other states, e.g., New York, have regulations for Subtitle D wastes that are significantly more stringent than minimum RCRA requirements and are even more stringent than Subtitle C waste regulations.

Regulations **in Germany** for liner systems, including closure and post-closure regulations, fall under the Waste Act (*Abfallgesetz AbfG*) and the Water Act (*Wasserhaushaltsgesetz WHG*). In addition, specific technical instructions have been promulgated under federal law. The objective is the establishment of a technical framework to reach the same degree of safety in containment, disposal, and management of waste materials in all German states. For the protection of the groundwater, there is the "1st. General Administrative Instruction on the Protection of the Groundwater for Storage and Deposition of Waste" (*1. Allgemeine Verwaltungsvorschrift zum Schutz des Grundwassers bei der Lagerung ung Ablagerung von Abfällen vom 31.01.1990, GMB1.S.74*). For hazardous waste and for municipal waste landfills, the federal government has issued: "Technical Instructions on the Storage, Chemical, Physical and Biological Treatment, Incineration, and Landfilling of Waste" (*TA Abfall 10.04.1990, GMBl. S. 170;* or **TA-A**), and "Technical Instructions on Recycling, Treatment, and other Management of Municipal Waste" (*TA Siedlungsabfall, 01.06.1993, Bundesanzeiger;* or **TA-SI**). In a complimentary manner, as in the United States, some of the German states have supplemented these federal regulations with their own state regulations. There are a number of specific regulations on waste management, issued by pertinent committees of the particular state involved.

Additionally, in Germany, there are government appointed task groups and professional groups which establish technical guidance. For example, the geotechnical aspects of solid waste landfills have been compiled by a task group of the German Geotechnical Society, *Deutsche Gesellschaft für Geotechnik*, and distributed as technical recommendations under the German title *Empfehlungen des Arbeitskreises "Geotechnik der Deponien und Altlasten;"* or *GDA*. The topics of the *GDA* recommendations were adapted to both international and European conditions and published as "Geotechnics of Landfill Design and Remedial Works; Technical Recommendations." The technical rules for the application of geomembranes in landfill engineering have been elaborated under the direction of the Federal Institute for Material Research and Testing, *Bundesanstalt für Materialforschung und-prüfung;* or *BAM*, and also published by the organization.

The parallelism between regulations in the United States and Germany can be seen in Table 1.1. The distinctions will become more apparent in the descriptive sections that follow.

Regulations for landfill liner systems (including covers for new and existing waste) in **countries other than the United States and Germany** vary from completely nonexistent to the levels prescribed in the preceding text. Indeed, most countries in the world are involved with, or concerned about, waste-containment systems. See, for example, Mackey (1996), who writes on the situation in Thailand.

TABLE 1.1. Parallel Regulation Categories in the USA and Germany

Country	Construction Debris/ Mineral Solid Waste	Municipal Solid Waste	Hazardous Waste	Abandoned Waste Dumps
United States	RCRA Subtitle D	RCRA Subtitle D	RCRA Subtitle C	CERCLA
Germany	Category I	Category II	Category III	Category II (generally)

It should be recognized, however, that although regulatory requirements are discussed in this book, there is no intention by the authors to provide complete coverage of regulatory requirements. The reader is advised that regulations change over time, and one should always consult with the appropriate authorities on any project to determine the current regulatory requirements. Regulatory requirements are presented in this book simply to give the reader a general idea of the requirements in the United States and Germany, which have (and will probably continue to have) a major influence on current design practices in North America and Europe, respectively, and in many other parts of the world.

1.2.1 Municipal Solid Waste Landfill Covers in the United States

The regulations dealing with final covers for municipal solid waste landfills in the United States are found in Title 40, Part 258, Subpart F (closure and post-closure care) of the Code of Federal Regulations; or CFR. The citation for the applicable regulation is thus 40 CFR 258.

The basic requirement as outlined in 258.60(a), states that:

"Owners or operators of all MSW landfill units must install a final cover system that is designed to minimize infiltration and erosion. The final cover system must be designed and constructed to:

1. have a permeability less than or equal to the permeability of any bottom liner system or natural subsoils present, or a permeability no greater than 1×10^{-5} cm/sec, whichever is less, and
2. minimize infiltration through the closed MSW landfill by the use of an infiltration layer that contains a minimum 450 mm of earthen material, and
3. minimize erosion of the final cover by the use of an erosion layer that contains a minimum 150 mm of earthen material that is capable of sustaining native plant growth."

Section 258.60(b) permits the director of an approved state to approve an alternative final cover design that includes an equivalent infiltration layer and erosion layer. Section 258.61 requires post-closure care and maintenance for at least 30 years, unless a different period is approved by the director of an approved state.

1.2.2 Hazardous Waste Landfill Covers in the United States

The regulations dealing with hazardous waste landfill and surface impoundment cover requirements in the United States are found in Title 40, Parts 264 and 265, of the Code of Federal Regulations (40 CFR 264 and 40 CFR 265). Part 264 deals with permitted facilities and Part 265 with interim-status facilities. Interim-status facilities are, in general, those facilities that were in existence on November 19, 1980. Three subparts of each of Parts 264 and 265 deal with general closure requirements: Subpart G—Closure and Post-Closure; Subpart K—Surface Impoundments; and Subpart N—Landfills. Each subpart contains several sections important to planning, design, and construction of covers.

Design objectives for final cover systems are stated in 264.228(a)(2)(iii) for permitted surface impoundments, in 264.310(a) for permitted landfills, in 265.228(a)(2)(iii) for interim-status surface impoundments, and in 265.310(a) for interim-status landfills. These sections require that the final cover be designed and constructed to provide for the following features.

> 1. Provide long-term minimization of migration of liquids through the closed surface impoundment or landfill.
> 2. Function with minimal maintenance.
> 3. Promote drainage and minimize erosion or abrasion of the cover.
> 4. Accommodate settling and subsidence so that the cover's integrity is maintained.
> 5. Have a permeability less than or equal to the permeability of any bottom liner system or natural subsoils present.

There are few differences between permitted and interim-status unit closure/post-closure regulations under Subpart G of Parts 264 and 265. The major difference is that, for interim-status units, public notice for changes to the approved closure and post-closure plans is not required. Changes to plans for permitted units require permit modifications which, in turn, require public notice and comment.

There are three significant differences between permitted and interim-status unit final cover regulations under Subparts K and N of Parts 264 and 265. Part 264.303 requires monitoring and inspection to ensure that geosynthetic and soil materials used in the cover are watertight and structurally uniform. Such a requirement is not included in Part 265 for interim-status units. The EPA recommends that a construction quality assurance (CQA) program be employed for covers being built at both permitted and interim-status units. The EPA believes that a site-specific CQA inspection program is necessary to ensure that cover design specifications are met.

A second difference in requirements is that, while leachate collection and removal activities are required after closure under 40 CFR 264.310, for permitted units, they are not required under Part 265 for interim-status units. The absence of a stated post-closure leachate collection and removal requirement makes cover performance for interim-status units even more important. It should be noted that, under the broader performance standards of 40 CFR 265.111, the EPA may still require leachate collection during post-closure at an interim site.

The third, and perhaps most significant, difference is in the requirements of 40 CFR 264.310(a)(5) and 40 CFR 265.310(a)(5). These subsections require that the cover have a permeability less than or equal to any bottom liner or natural subsoil present. For interim-status units, without an engineered liner, the cover could presumably be of relatively permeable materials, but the EPA may impose the standards of 40 CFR 265.111 and require a more impermeable cover.

For permitted landfills to meet the requirements of 40 CFR 264.310, the cover must have a permeability no greater than that of the double liner required under 40 CFR 264.301. The EPA has not interpreted this to mean that the final cover for a permitted unit must actually contain a double liner. Rather, the final cover should include a layer whose liquid-rejection performance is equal to or better than the bottom composite liner (i.e., a geomembrane in full contact with a low-permeability compacted clay or an equivalent clay barrier) of the double-liner system. In all cases where a geomembrane is used in the bottom liner, a geomembrane or equivalent liner should also be used in the cover. Similarly, a hydraulically equivalent material to the minimum 900 mm of compacted clay having a hydraulic conductivity of 1×10^{-7} cm/s or less is required in the final cover system, if such a layer underlies the geomembrane in the bottom liner system. This does not mean, however, that exactly the same barrier materials have to be used in both the liner and cover. In fact, just the opposite is true. For example, different geomembrane materials of equivalent performance may be used, with a geomembrane having excellent chemical resistance characteristics selected for the bottom liner and a geomembrane with excellent flexibility and elongation characteristics selected for the cover system. Also, geosynthetic clay liners (GCLs) may be used instead of compacted clay liners (CCLs) if their hydraulic performance can be shown to be equivalent to 900 mm of compacted clay having a hydraulic conductivity of 1×10^{-7} cm/s or less.

The EPA also recommends using the composite geomembrane/clay barrier in interim-status unit covers. While 40 CFR 265.310(a)(5) might allow a less effective design, the long-term protection from infiltration provided by a composite barrier justifies its use for all hazardous waste units.

1.2.3 Covers for Abandoned Dumps and Remediation Action Projects in the United States

The U.S. regulations dealing with abandoned dumps and other remediation projects (often referred to as "Superfund" projects) are found in Title 40, Part 300, of the Code of Federal Regulations (40 CFR 300). Subpart E ("Hazardous Substance Response") provides information on remedial site evaluation (40 CFR 300.420), establishing remedial priorities (40 CFR 300.425), remedial investigation/feasibility study and selection of remedy (40 CFR 300.430), and remedial design/remedial action, operation, and maintenance (40 CFR 300.435).

The overall goal of remediation as stated in 40 CFR 300.430(a) is to implement remedies that eliminate, reduce, or control risks to human health and the environment, and that maintain protection over time. As stated in 40 CFR 300.430(a)(iii), the EPA expects to use treatment to address the principal threats posed by a site, wherever practicable. Principal threats for which treatment is most likely to be appropriate include liquids, areas contaminated with high concentrations of toxic compounds, and highly mobile materials. The EPA expects to use engineering controls, such as containment, for waste that poses a relatively low long-term threat or where treatment is impractical. A large abandoned dump (landfill) that requires action under Superfund is an example of a site for which complete treatment is usually impractical and where engineering controls such as final cover systems are usually part of the remedial design. It is often appropriate to use a combination of methods, e.g., removal and treatment of liquids, and containment of solids and other residuals or untreated wastes. As stated in 40 CFR 300.430(a)(iii)(E), the EPA will consider using innovative technology when such technology offers the potential for comparable or superior treatment performance or implementability, fewer or lesser adverse impacts than other available approaches, or lower costs for similar levels of performance than demonstrated technologies.

No specific requirements are established for final cover systems in Superfund. Instead, criteria for evaluating alternatives for treatment or containment are established (40 CFR 300.430(a)(2)(e)(9). The EPA requires that a detailed analysis be conducted on the limited number of alternatives that represent viable approaches to remedial action after evaluation in the screening stage. If a final cover system is being considered as a containment alternative (either alone or in combination with other containment actions and/or with treatment alternatives), then the cover system should be considered in light of EPA criteria. The EPA requires that the analysis of alternatives under review must reflect the scope and complexity of site problems and alternatives being evaluated. Nine criteria are to be considered:

1.   Overall protection of human health and the environment. Alternatives must be assessed to determine whether they can adequately protect human health and the environment, in both the short- and long-term, from unacceptable risks by eliminating, reducing, or controlling exposures to levels established during development of remediation goals. For a final cover system, specific performance criteria (e.g., allowable amount of water percolation through the cover) may be developed if they are linked to risk management and the overall remediation goals.

2.   Compliance with ARARs. The alternatives must be assessed to determine whether they attain Applicable or Relevant and Appropriate Requirements (ARARs) under state or federal environmental laws or regulations.

3.   Long-term effectiveness and permanence. Alternatives must be assessed for the long-term effectiveness and permanence they afford, along with the degree of certainty that the alternative will prove successful.

4.   Reduction of toxicity, mobility, or volume through treatment. The degree to which alternatives employ treatment that reduces toxicity, mobility, or volume must be assessed.

5.   Short-term effectiveness. The short-term effectiveness of alternatives must be assessed, including short-term risks that might be posed to a community during implementation, potential impacts on workers during remedial action, and time until protection is achieved.

6.   Implementability. The implementability must be assessed considering technical feasibility, administrative feasibility, and availability of services and materials.

7.   Cost. Costs must be assessed, including capital, operating, and maintenance costs.

8.   State acceptance. The state's comments must be considered.

9.   Community acceptance. The community's acceptance of various components of the remedial alternatives must be considered.

1.2.4 Solid Waste Landfill Covers in Germany

In Germany, landfills of Categories I, II, and III are distinguished from one another as follows:

- Category I landfills consist of mineral solid waste having low pollution potential. Such landfills are not anticipated to undergo chemical or biological reactions.
- Category II landfills are municipal solid waste landfills and currently receive degradable materials. However, according to the Federal Instructions (TA-SI), by the year 2005, the maximum allowable carbon content of the waste must not exceed 5%. Thus, waste received at a landfill at that time will be mainly incinerator ash. Until then, Category II landfills are essentially bioreactors whereby settlements and gas generation must be considered in the cover system.
- Category III landfills consist of hazardous wastes. The waste must be deposited such that no major volume changes will occur over time. Thus long term settlements of the cover system must be minimized.

German cover regulations regarding the above three categories of landfills are focused on surface/protection layers, drainage layers, barrier layers, and gas venting/foundation layers. Each will be explained within the context that the ultimate objectives of the final cover system are to provide the following:

- Prevention of infiltration of water into the underlying waste.
- Prevention of gas emissions from the underlying waste.
- Promotion of growth of vegetation on the upper surface.
- Facilitation of landscaping of the site to provide a reasonable appearance.

1.2.5 Mineral Solid Waste Landfill Covers in Germany

These waste masses have low pollution potential and are not expected to have significant long-term settlement of the cover system.

The **surface/protection layer** must be adequate for long-term maintenance and reliability.

The **drainage layer** above the barrier, according to the German regulations (TA-A and TA-SI), is a layer of granular soil at least 300 mm thick with a minimum hydraulic conductivity of 0.1 cm/s. At many sites, the same 16/32 mm rounded stone required in the leachate collection system is used in the cover system as well. If so, and if a geomembrane is in the underlying barrier system, puncture protection must be provided. As a result, geosynthetic drainage materials are receiving greater usage. Geonets, as commonly used in the United States, are not as common in Germany. Instead, drainage geocomposites made from stiff filaments, drainage cores, and other variations are used. The difference in materials used in the United

States and Germany are primarily the result of historic differences in manufacturing approaches rather than technical performance issues. The drainage layer must be designed within the context of German regulations. The minimum gradient is 5% and the maximum slope should not be steeper than 3(H) on 1(V). This can be exceeded on a site-specific basis, however, stability becomes the major concern. Geogrid inclusions have been used for slope reinforcement at some older German landfills, which have been as steep as 2(H) to 1(V). The primary drainage layer design criteria are adequate hydraulic flow capacity, stability within the product (if it is a geocomposite) and adequate shear strength of all interfaces involved.

The **hydraulic barrier layer** in Category I landfill covers, according to German regulations, consists of compacted clay placed in two 250 mm thick lifts, with a hydraulic conductivity of 5×10^{-7} cm/s or less. Alternatively, a GCL can be used if shown to be technically equivalent. To aid in the assessment of technical equivalency, the German Institute of Construction Technology (DIBt), together with an independent group of technical experts, has established criteria. Decisions are pending.

Category I landfills are not anticipated to have gas generation, therefore, a **gas venting layer** is not required.

The **foundation layer** over the waste is critical in setting the grade for all of the overlying cover layers, since each of them is constructed in a sequential manner.

1.2.6 Municipal Solid Waste Landfill Covers in Germany

As described previously, this category of German landfills is in a transition (to be completed by the year 2005) from accepting degradable municipal solid waste to eventually being mainly incinerator ash from combusted household waste materials. The major differences between the two types of waste are the amount of gas produced and the accompanying settlement over time.

Regarding the **surface, protection, and drainage layers** over the barrier layer, the situation is the same as described for Category I landfills.

The German regulations (TA-SI) do not give detailed requirements for the **hydraulic barrier layer** of old Category II landfill covers. Therefore, most barriers are similar to Category I, i.e., a 500 mm thick compacted clay layer of 5×10^{-7} cm/s, or lower, hydraulic conductivity. GCLs are also being considered for these landfills.

The **hydraulic barrier layer** for Category II landfills will eventually consist of a geomembrane over compacted clay. The compacted clay is described above. The geomembrane, by German TA-SI regulations, must be high-density polyethylene (HDPE) with a thickness of at least 2.5 mm. The regulations permit the use of recycled polymers, however, none have been approved to date. The geomembrane must be approved by the Federal

Institute for Material Research and Testing, *Bundesanstalt für Matertial-forschung und—prüfung* (BAM). Textured geomembranes have been used, but they must also be BAM approved. Particular attention is paid to long-term strength and stress cracking resistance. If slope stability analysis suggests a concern, geogrid reinforcement can be used to provide an adequate factor of safety against sliding.

A **gas venting** layer must be provided in current Category II landfills accepting degradable municipal solid waste. When using a natural soil system, it must be adequately permeable, have less than 10% calcium carbonate content, and be provided with a network of collection pipes of at least 100 mm diameter. Alternatively, a geosynthetic gas venting layer can be provided. The gas that is generated by the degrading waste must be collected and transmitted to a suitable container or utilized for energy production. For future Category II landfills where no gas is generated, a gas venting layer will not be necessary.

The soil **foundation layer** placed directly over the waste is the same as described previously.

1.2.7 Hazardous Waste Landfill Covers in Germany

Category III landfill wastes are considered to be hazardous waste landfills in the German TA-A and TA-SI regulations.

Regarding the **surface, protection and drainage layers** over the barrier layer, the situation is the same as described in Category I and II landfills.

The **hydraulic barrier layer** must be a geomembrane/compacted clay liner composite system. Details of both materials were described in the Category II discussion describing the future class of incinerated municipal solid waste. The notable exception is that the hydraulic conductivity of the compacted clay liner component beneath the geomembrane must be an order of magnitude lower, i.e., the hydraulic conductivity of the compacted clay liner must be equal to 5×10^{-8} cm/s or less. It is important to note that the TA-A regulations require a leak detection system. In some cases, double geomembrane liners with intermediate leak detection layers have been constructed. This, however, does not fit the German philosophy of having only low permeability layers with the hydraulic barrier system. Alternatively, some type of continuous monitoring system is possible; a number of which are emerging (Stief 1996).

A **gas venting** layer is not necessary under German regulations, and the **foundation layer** is a site-specific situation. Compaction and placement must be adequate to facilitate the placement of the compacted clay liner above.

1.2.8 General Comments on Cover Requirements in Germany

The German regulations (TA-A and TA-SI) allow for material substitutions, provided the replacement material is fully justified, e.g., a geosynthetic clay liner for a compacted clay liner. Furthermore, they suggest large-scale test plots for introduction of materials new (or different) from those stipulated in the regulations.

There are no special German regulations for the design and construction of covers for abandoned dumps. According to the assessment based on state regulations, cover systems of abandoned dumps are constructed in compliance with the rules for engineered landfills.

Great emphasis is placed in the German regulations on both manufacturing quality assurance (MQA) and construction quality assurance (CQA). Currently, all types of geosynthetics (geomembranes, geocomposite drains, geotextiles, geosynthetic clay liners, and geogrids) are being used in association with natural soils for landfill capping systems. Thus both MQA and CQA are emphasized just as they are in U.S. regulations. While differences do occur between the German and United States regulations, it is encouraging that these two countries do cooperate and share their separately developed rules and regulations. See Stief (1986) and Gartung (1995 and 1996) for additional details on both the liner and the cover systems for German landfills.

1.2.9 Summary Comments on Regulations

German and United States regulations have similarities and differences. The main similarities are the specification of the same types of components for multi-layered cover systems and similar requirements for those components. The main differences are found in the details of specific requirements and in the more stringent specification approach in Germany (e.g., required approval of materials and strict requirements for manufacturing quality control). Many people criticize U.S. regulations for being too specific and detailed, but they are not the most stringent regulations in the world. German regulations are. Yet the concern expressed in the United States that excessively prescriptive regulations tend to stifle innovation and creativity also rings true in Germany.

The challenge for engineers and regulators is not to focus energy on complaining about regulations, but, instead, to concentrate on building a consensus for different approaches that may be desirable. All regulations allow alternatives, and the key is to continue to search for better alternatives.

1.3 LIQUIDS MANAGEMENT PRACTICES

Liquids that arrive with the incoming waste materials, augmented by meteoric water in the form of rainfall and snowmelt, commingle with the breakdown products of the waste constituents to create "leachate." As the leachate flows gravitationally to the base of the landfill, it gradually takes on the chemical characteristics of the waste. In modern landfills, the leachate is prevented from becoming fugitive into the local groundwater or surface water by the barrier of the bottom liner system. Immediately above the bottom liner system is a leachate collection and removal system.

There are two approaches widely practiced for management of leachate generation. One approach is to try to minimize the amount of liquid percolating through the cover and into the underlying waste (the so-called "dry tomb landfill"). The other approach is to intentionally promote water infiltration so that accelerated biodegradation of the waste occurs. These approaches are briefly discussed in the following sections. Of course, for old or abandoned landfills without a leachate collection and removal system, there is no option for liquids management other than construction of leachate extraction wells or trenches.

1.3.1 Standard Leachate Withdrawal

Since the early 1980's, the standard leachate management strategy at lined landfills has been to withdraw the leachate as it flows through the waste and eventually arrives at the down-gradient manhole or sump area. In many regulations, the maximum leachate head allowed on the liner system is 300 mm. The leachate collection system must be designed accordingly. Leachate removal is generally by means of a vertical manhole within the waste or an inclined sidewall riser. Both schemes are illustrated in Figure 1.4. The collected leachate is then treated and discharged in an environmentally safe and secure manner. The practice of complete gravity flow of leachate, whereby a bottom liner penetration is required, is technically allowed but seldom practiced due to the difficulty of making a leakproof penetration at the most sensitive location in a landfill.

The quantity of leachate that is generated is both site-specific and waste-specific. The quantity is estimated from predictive water balance analyses, such as described in Chapter 4. The quantity depends upon a number of factors, among which are the following:

- local site hydrology,
- landfill-specific geometry and surface flow pattern,
- type of the solid waste,
- age of the waste,
- moisture content of the solid waste, and
- thickness of the solid waste.

To give an idea of the quantity that might be involved, New York State has completed a survey of its landfills which indicates an average of 25,000 to 30,000 L/ha-day of leachate must be removed. A thorough review of the subject can be found in Chapter 4.

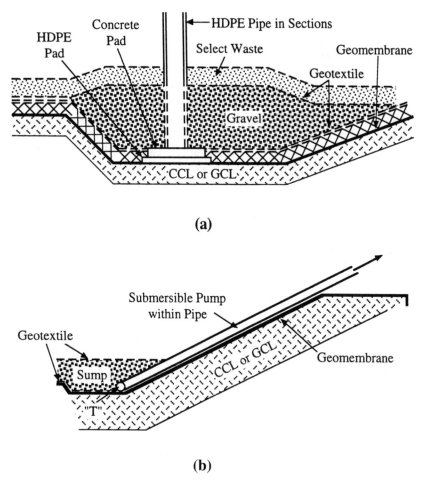

FIG. 1.4. *Methods for Removal of Leachate from the Base of Landfill*

When the landfill reaches its permitted height, either a temporary or permanent cover is placed over the waste. Once the waste is covered, the quantity of leachate generated greatly diminishes. Over a period of many years, leachate flow eventually decreases, often to a negligible amount.

After waste placement in a landfill and covering, the waste (if putrescible) degrades and settles. The manifestation of this degradation and the accompanying surface settlements are:

- changing characteristics of the leachate (sometimes characterized by high suspended solids and high microorganism content),

- liberation of gases (some lighter than air, some heavier), and
- subsidence of the surface of the waste on which the final cover has been, or must eventually be, placed.

This subsidence evidences itself as total settlement of the entire cover, and possibly localized differential settlement as well. The cover system should be designed to withstand the anticipated total and differential settlement. Edgers et al. (1990), König et al. (1996), and Spikula (1996) have accumulated data for municipal solid waste landfills indicating total settlements up to 30% of the height of the landfill extending beyond 20 years (see Figure 1.5). Presumably the liquids management strategy (if any) in these case histories was standard leachate withdrawal. Using this data and estimating how other MSW liquids management strategies might influence settlement, Table 1.2 is offered by the authors. Note that a range of estimates for differential settlement is also included in Table 1.2. Clearly, this latter topic deserves closer observation than it has had in the past, since it becomes a dominant issue in the design of the barrier layer in a final cover system.

1.3.2 Leachate Recirculation

Regulations in the United States, promulgated in 1993, allow for an alternative to the procedure just described for managing leachate at municipal solid waste landfills. It is called *leachate recirculation*. Leachate is

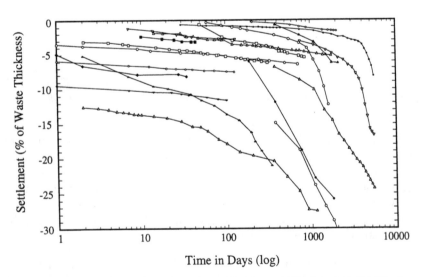

FIG. 1.5. *Total Settlement Data from Number of Municipal Solid Waste Landfills [After Edgers et al. (1990); König et al. (1996); Spikula (1996)]*

TABLE 1.2. Impact of Liquids Management Practice on Cover Settlement at Municipal Solid Waste Landfills[1,2]

Leachate Management Practice	Total Settlement		Differential Settlement[3]	
	Amount	Time	Amount[4]	Time
Standard leachate withdrawal	10–20%	≤30 yrs.	Little to moderate	≤20 yrs.
Leachate recirculation	10–20%	≤15 yrs.	Moderate to major	≤10 yrs.
None, e.g., at abandoned landfills or dumps	Up to 30%	>30 yrs.	Unknown	>20 yrs.

[1]Hazardous waste landfills, industrial waste landfills, and incinerator ash monofills usually have much less settlement than the amounts listed in this table.
[2]The estimates in this table regarding the impact of the liquids management practice on settlement of landfill covers are based on sparse data. They are meant to be a guide only, and site-specific estimates are required to develop more realistic figures for any particular final cover project.
[3]The estimates in this table regarding differential settlement amount and time are also based on very sparse data. Clearly, field monitored data is needed in this regard.
[4]These qualitative assessment terms are also affected by the density of the waste; well-compacted waste produces less differential settlement than poorly-compacted waste.

withdrawn at the removal sump or manhole and then reintroduced back onto or into the waste.

The leachate recirculation concept is meant to be continuous and is intended to accelerate the degradation of the solid waste. The leachate may be treated as it is removed and repeatedly reinjected into the waste. A U.S. EPA workshop on leachate recirculation described the mechanisms involved and presented details on the various implementation strategies (EPA 1996).

Leachate recirculation would probably cause total settlement to occur more rapidly and possibly to a greater amount than the standard leachate management practice involving withdrawal, treatment, and off-site discharge. Leachate recirculation may also result in greater differential settlement, but this is not known to be the case. Table 1.2 provides suggested estimates for differential settlement. As of 1997, there are approximately 30 landfills that practice leachate recirculation in the United States that are being monitored. Leachate recirculation, however, is far from new. Landfill owners throughout the world have used the landfill itself for temporary leachate storage during times of inadequate storage capacity or treatment plant shut down and/or over loading. In Italy, the method has been practiced routinely for many years.

1.3.3 Leachate Within Abandoned Dumps

Abandoned dumps, which are essentially nonengineered facilities, can be characterized as those in which there has been essentially no liquids management of any type practiced. Indeed, most abandoned dumps have neither a leachate collection system nor ancillary sumps or manholes with which to remove the leachate. Thus, the leachate is contained in the waste, unless it becomes fugitive into the surrounding environment, which is possible since abandoned dumps usually have no engineered bottom liner system. This, of course, accounts for the concern over abandoned dumps and their threat to the surrounding environment. In this case, vertical cutoff walls, pump-and-treat systems, and other remediation methods are generally implemented (Rumer and Mitchell 1996).

With abandoned landfills, the potential settlement of the cover system is uncertain because the nature and thickness of the waste mass is largely unknown. Because settlements, both total and differential, are difficult to estimate, conservative estimates should be made insofar as a cover design is concerned.

1.4 GENERAL COMPONENTS OF COVER SYSTEMS

Final cover systems for municipal solid waste landfills, hazardous waste landfills, and corrective action projects (including abandoned dumps) have a number of very similar elements. As shown in Figure 1.3, there are six basic components of a final cover system: (1) surface layer; (2) protection layer; (3) drainage layer; (4) hydraulic/gas barrier layer, (5) gas collection layer; and (6) foundation layer. Not all components are needed for all final covers. For example, a drainage layer might not be needed for a cover system located in an arid region. Similarly, a gas collection layer may be required for some covers but not others, depending upon whether the waste or contaminated material is producing gases that require collection and management. In addition, some of the above listed layers may be combined. For instance, the surface layer and protection layer are often combined into a single layer of soil that forms the "cover soil." Likewise, the gas collection layer (when made from sand) is often combined as a single layer with the foundation layer. The materials that are typically used for the various components are noted in Figure 1.3. The critically important hydraulic barrier layer often is a composite consisting of two components, e.g., a geomembrane overlying a clay liner, where the clay is either a compacted clay liner or a geosynthetic clay liner.

1.4.1 Surface Layer

The most commonly used material for the surface layer is fertile topsoil. Such vegetated topsoil helps to minimize erosion and promote transpiration of water back to the atmosphere. Vegetation also creates a leaf cover above the soil to reduce rainfall impact and decrease wind velocity on the soil surface. The main drawbacks to a surficial layer of soil are problems in initiating vegetative growth (particularly during times of the year when conditions for plant growth are unfavorable) and vulnerability to wind or water erosion if the vegetative growth is not suitably thick, e.g., following a period of drought. Often, topsoil is the most costly component of a cover system due to high demand and low availability.

Erosion can be controlled temporarily by placing a geosynthetic erosion control material at the surface. Such materials are commonly used on bare slopes next to highways, and hold the soil in place until vegetation can become established.

At some sites, cobbles have been used as the surface layer. Cobbles are highly resistant to erosion from both wind and water. A problem with cobbles, however, is that they allow water to enter but do little to promote evapotranspiration back to the atmosphere. Thus, a surficial layer of cobbles may allow far greater infiltration of water into the cover system than does vegetated topsoil.

Paving materials, e.g., asphaltic concrete, have occasionally been used for the surface layer, but paving materials are not normally recommended for permanent covers because of possible settlement-induced cracking and because ultraviolet radiation will eventually oxidize and degrade the exposed material. Other types of flexible "hard-armor" materials are possible, e.g., articulated blocks and concrete filled geotextile mattresses, and will be described later.

1.4.2 Protection Layer

The protection layer may serve one or more of the following functions:

> 1.   Store water that has infiltrated into the cover until it is later removed by evapotranspiration;
> 2.   Physically separate the underlying drainage and barrier layer components and buried waste from burrowing animals and plant roots;
> 3.   Minimize the possibility of human intrusion into the contaminated material;
> 4.   Protect underlying layers from excessive wetting/drying (which could cause cracking of fine-grained soils);
> 5.   Protect underlying layers in the surface barrier from freezing (which could cause frost heave of underlying soils or cracking of fine-grained soils).

Locally available native soil is the material that is most commonly used for the protection layer. Sometimes the surface and protection layers are constructed of the same soil material and are combined into a single "cover soil" layer. Soil can be very effective in storing water for later evapotranspiration and it provides protection against freezing of underlying materials if the thickness of the protection layer is sufficient.

Cobbles within the protection layer have occasionally been considered as a barrier to plant roots and burrowing animals. A layer of cobbles has been considered as a biointrusion protection layer for radioactive waste disposal sites. The reason why a layer of cobbles is not normally considered for other types of waste containment (e.g., municipal or hazardous waste landfills) is that uptake of dangerous compounds by plant roots and burrowing have not been known to be problems in those facilities.

If the protection layer is placed directly on a barrier layer, a plane of potential slippage exists at their interface. The risk of instability will be particularly acute following prolonged periods of rain if no drainage layer is provided beneath the protection layer. The design engineer must ensure that an adequate factor of safety against slippage exists at this and all other interfaces in the final cover system. If the originally designed cover does not have an adequate factor of safety, the steps that are usually taken to increase the safety margin include use of different materials (stronger soils or materials such as textured geomembranes with higher strength along the interface), addition of a drainage layer, flattening of slopes, shortening of slope lengths with intermediate berms, or reinforcement of cover soils with geogrids or high strength geotextiles. Slope stability is covered in Chapter 5.

1.4.3 Drainage Layer

A drainage layer is often placed below the protection layer and above the barrier layer. There are three reasons why a drainage layer might be desirable:

> 1. To reduce the head of water on the barrier layer, thus minimizing infiltration;
> 2. To drain the overlying protection layer, thereby increasing its water storge capacity; however, this can also have a deleterious effect if it dries the overlying soil to the point where vegetation fails to thrive;
> 3. To reduce and control pore water pressures in the cover soil, and thus improve slope stability.

The third reason cited above is often the most important. In areas that receive sufficient rainfall to soak the protection layer to a significant depth, a drainage layer is usually essential to maintain stability of slopes. If the cover soil becomes saturated and the water table rises to the surface of the cover system, the factor of safety against slope failure reduces to about half the value for a nonsaturated cover. In arid areas or on relatively flat portions of final cover systems, a drainage layer may or may not be needed, depending on site-specific requirements.

Selection of the drainage material type is usually based on availability of local materials, economics, and design life. The most commonly used material is sand, although gravel is sometimes used if a high hydraulic transmissivity is needed or if gravel is more abundant in the region. Geosynthetics (geonets and geocomposites) are also frequently used, and alternative materials such as shredded tires have occasionally been used.

The drainage layer will usually underlie the soil protection layer. Adequate filtration must be provided so that finer particles from the overlying protection layer will not migrate into the drainage layer and cause excessive clogging. The filter layer could consist of soil, but is more frequently a geotextile.

If the drainage layer contains large objects (e.g., stones) or sharp objects (e.g., reinforcing wire in shredded tires), there is a risk of puncture of an underlying geomembrane. If puncture is a problem, a cushion layer between the drainage layer and underlying barrier may be needed. It can be soil, but more often it is a relatively thick nonwoven needle-punched geotextile.

Controlling the discharge of water from the drainage layer is an important detail. Water must be allowed to discharge freely from the drainage layer at the base of the cover system, through what is often called a "toe drain." If the toe drain excessively clogs, freezes, or is of inadequate capacity, the toe of the slope will become saturated, develop upgradient excess pore water pressure, and potentially become unstable. Drainage pipes within the toe must have adequate capacity and freeze protection. Adequate filtration around the toe drain is crucial.

1.4.4 Hydraulic/Gas Barrier Layer

The barrier layer (often called "hydraulic" barrier layer) is generally viewed as the most critical component of an engineered final cover system. The barrier layer minimizes percolation of water through the cover system directly by blocking water and indirectly by promoting storage or drainage of water in the overlying layers, where water is eventually removed by runoff, evapotranspiration, or internal drainage. Furthermore, the barrier layer prevents landfill gases from escaping into the atmosphere. Such gases have been shown to be a major source of air pollution and ozone depletion. Recognize, however, that the barrier layer forces the gas to migrate beneath it and must be used in conjunction with an effective landfill gas venting or collection system to control lateral gas migration.

The barrier layer for hazardous waste disposal facilities has typically consisted of a composite layer made up of a geomembrane overlying a compacted clay liner (CCL), the latter having a thickness of 600 mm and a hydraulic conductivity $\leq 1 \times 10^{-7}$ cm/s. Recall that the German regulations call for $k \leq 5 \times 10^{-7}$ cm/s in Category I and II landfill covers and $k \leq 5 \times 10^{-8}$ cm/s in Category III landfill covers. In recent years, an alternative to the compacted clay liner, i.e., a geosynthetic clay liner (GCL), has been considered, found to be technically equivalent in many cases, and used accordingly.

Occasionally, other types of materials are used. A sprayed-on asphaltic membrane has been employed in some cover systems as an alternative to a geomembrane for site remediation projects. Asphaltic concrete has also been used as a substitute for a compacted clay liner. Asphaltic concrete is not vulnerable to desiccation cracking, which makes it an attractive alternative for some facilities located in arid regions. However, oxidation of asphalt can eventually cause embrittlement and sensitivity to settlement of the cover system. Barrier layer materials, along with their relative advantages and disadvantages, are discussed in detail in Chapter 2.

Perhaps the most critical issue with respect to the barrier layer is its protection. All of the barrier materials discussed above can provide extremely high impedance to downward percolation of water if they are properly installed. Similarly, all of them can be compromised by desiccation, puncture, or other types of damage. Compacted clay liners (CCLs) are especially vulnerable because they have low resistance to wet-dry cycles (which causes desiccation cracking), freeze-thaw cycles (which increases hydraulic conductivity), and distortion caused by differential settlement (which can cause tensile cracks to form). Thin liners, such a geomembranes, geosynthetic clay liners, and sprayed-on asphalt, are vulnerable to accidental puncture, but an occasional puncture is not nearly so damaging in a final cover system as it can be at a more critical location, such as the sump in a bottom liner.

The barrier layer is also intended to stop the upward migration of gases, if they are present. Some materials are better barriers to gas movement than others. Some types of geomembranes, for instance, can be highly impermeable to gases, while the gas permeability of compacted clay liners is extremely sensitive to the water content and degree of cracking in the CCL. A saturated, crack-free CCL would make an excellent barrier to gas migration, but a relatively dry or significantly cracked CCL would make a very poor barrier.

The use of thin liners, such as geomembranes, geosynthetic clay liners, or sprayed-on asphalt, creates a potential problem with interface shear between the thin liner and both underlying and overlying components. The potential for interface shear should be addressed during the design process by determining interface shear parameters for site-specific materials and by use of appropriate methods of slope stability analysis. The use of published values of interface shear without confirmation for the project-specific materials during the design or construction phase is inappropriate because of the inherent variability of both geosynthetic and natural materials. Chapter 5 will discuss slope stability issues in detail.

1.4.5 Gas Collection Layer

The purpose of a gas collection layer is to collect gases from the decomposition of putrescible wastes or volatile organics from wastes containing organic chemicals. The gas collection layer may be constructed of sand, gravel, geonet, geotextile, geocomposite, or other gas-transmitting material. Gas is captured in the gas collection layer and flows from it into periodically-spaced collection pipes or vents. Flow of gas may occur passively under the natural pressure gradient within the closed landfill or may flow actively through assistance from a vacuum system located at the surface. A gas collection layer is necessary for those solid waste materials that produce gas or volatiles which must be subsequently treated or released in a controlled manner. This generally includes municipal solid waste landfills, abandoned landfills, and some miscellaneous landfills.

The gas collection layer must have a high in-plane transmissivity and must not become excessively clogged with fine-grained materials. Filters may be needed to prevent fine materials from adjacent layers (above and/or below) from excessively clogging the gas collection layer.

In some cases, where the waste is of such character that vertical migration of gases is impeded, full-depth vent structures to the bottom of the waste mass may be needed. These structures are designed to prevent the horizontal migration of gases out of the landfill into the surrounding soil. Active rather than passive systems may be required in some cases to adequately remove accumulated gases. Indeed, vents (when properly designed) can act as an alternative to a complete gas collection layer.

1.4.6 Foundation Layer

The lowest layer in the cross section of a multi-component final cover system is a foundation layer. The foundation layer is often the interim cover that has already been placed, perhaps with some additional compaction before the final cover is constructed. If there is no interim cover, if the interim cover is too thin, or if differential settlement has created an irregular surface, then additional soil is typically brought in, spread, and compacted to form the foundation layer. If the foundation layer is sand or gravel, it may be considered to be the gas collection layer, but this is a site-specific situation.

1.5 GENERAL CONCERNS

This section is written to review some of the more important concerns that have lead to past problems with respect to final covers. The major causes of poor performance of final covers that the authors have observed are the following:

1. settlement/subsidence
2. slope instability
3. inadequate filtration
4. improper gas management
5. long-term erosion
6. unacceptable final use and aesthetics

Additional detail on each of these topics will be presented in subsequent chapters.

1.5.1 Settlement/Subsidence

All solid waste masses will settle over time, causing a subsidence of the upper surface and with it the engineered cover as well (recall Table 1.2). Many variables affect the magnitude of settlement, including the liquids management practiced at the specific landfill (recall section 1.3). Design estimates of both total settlement and differential settlement are required in order to analyze the long-term functioning and stability of the final cover.

In general, the upper surface of the solid waste mass will have a layer of soil placed over it acting as a foundation layer for the eventual placement of the cover materials. This foundation layer will normally be rolled with compactors. In some instances, deep dynamic compaction may be utilized to reduce the post construction settlement of the underlying waste mass (Galenti 1991).

The final grade of the foundation layer must take into account the estimated total settlement after cover placement. As was seen in Table 1.2, this can amount to from 10% to 20% of the thickness of the waste for municipal solid waste landfills. For industrial waste landfills and monofills, the settlement will generally be much lower. In the case of abandoned landfills, due to the unknown nature of the past waste disposal practice, estimates should be made very conservatively. Because of the uncertainties of estimating total settlement and/or differential settlement, a post-closure monitoring and maintenance plan for repair is strongly recommended.

Grading of the soil foundation layer to accommodate total settlement can take many forms. For example, past designs have been as follows:

- continuously crowned surface with drainage to the perimeter of the site,

- uniformly or incrementally sloping surface with drainage to one or two sides of the site,
- undulating (accordion) surface with drainage to one side of the site, and
- nonuniform crowned surface with benches forming drainage spirals with intermittent letdown channels for surface runoff.

Within the context of site-specific constraints and the estimate of total settlement, the above types of grading layouts are within the state-of-the-practice. By comparison, differential settlement is the most troublesome. There are many issues concerning differential settlement of solid waste. The most frequently asked questions are:

1. Will differential settlement occur?
2. If it occurs, what will be the extent and dimensions of the differential settlement void or trough?
3. Can surface grading be provided so that the water infiltrating through the cover soil does not form a "bathtub" in the barrier system?

In the situation where differential settlement is likely to occur and the localized depressions cannot be eliminated by suitable grading, the choices are (a) continuously grading and maintaining the site, or (b) reinforcement of the cover system. Such reinforcement can be provided by geogrids or high-strength geotextiles placed beneath the barrier system. The critical design parameter is the dimension of the anticipated differential settlement void or trough. Since it is generally not possible to predict where such differential settlement will occur, the reinforcement is usually placed over the entire landfill. The cost of such reinforcement is far from trivial, but nevertheless, reinforcement can be cost effective in light of the cost of continuous maintenance. Clearly, there is a pressing need for close-in surveys of solid waste covers to determine if, where, and how much differential settlement occurs in closed landfills of all types.

1.5.2 Slope Instability

Final covers are almost always sloped to maximize waste volume and to promote runoff. Sometimes, slopes are very steep, approaching grades of 50%. (The "grade" is defined as the vertical distance divided by horizontal distance, and is numerically equal to the tangent of the slope angle). The introduction of a geomembrane, geosynthetic clay liner, and/or compacted clay liner on a landfill cover slope steeper than a grade of 5 to 10% must be done with the utmost care. The design of slopes is within the state-of-the-practice, yet slides continue to occur, attesting to the difficulty of many situations and, in some cases, to the mistakes of the designer. Boschuk (1991) has investigated a number of such slides.

There are three dominant driving forces leading to slope instability: gravitational, seepage, and seismic forces. The *gravitational* forces, including live loads from construction equipment, are generally evaluated using a limit equilibrium method. Numerous design charts and computer codes are available. Backfilling should proceed up the slope, not downward. If it is necessary to backfill down the slope, the dynamic force of the construction equipment should be considered in the analysis. The adverse impact of *seepage* on slope stability is usually minimized or eliminated by using an adequate drainage system above the barrier layer. This drainage system conducts flow to a suitable outlet at benches or at the toe of the cover. Unfortunately, this objective is not always achieved, since most failures of cover slopes appear to be (at least in part) seepage induced. Some causes have been as follows:

1.   Inadequate drainage capacity at the toe of the slope where water accumulates.

2.   Fine sediments from the drainage material accumulating at the toe of the slope, thereby decreasing the ability of the granular material to drain.

3.   Fine, cohesionless, soil particles migrating through the filter (if one is present) and the drainage layer and then accumulating at the toe of the slope, thereby impeding drainage.

4.   Freezing of the drainage layer at the toe of the slope while the top of the slope thaws, thereby blocking drainage.

Seismicity is usually evaluated by a two-step process. First, a pseudo-static analysis is performed, and if the factor of safety of any representative cross section falls below unity, a deformation analysis is performed. The maximum computed deformations are then evaluated in terms of probable performance of the final cover system.

For all cases where the factor of safety is too low (for gravitational, seepage, and/or seismic forces), a redesign is required. Generally a factor of safety of at least 1.5 is the recommended minimum value for static loading of final covers. Approaches that are routinely used to increase the factor of safety are as follows:

1.   Flatten the slopes.

2.   Interrupt long slope lengths with intermittent benches.

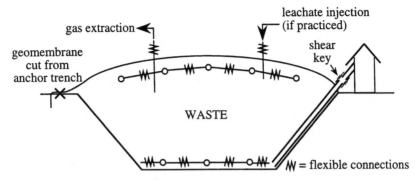

FIG. 1.6. *Concern Areas of Landfills with Respect to Seismic Activity and Possible Preventative Measures to Avoid Problems*

    3.   Construct reinforced earth berms or buttress walls at the toe of the slope.

    4.   Include veneer reinforcement (geogrids or geotextiles) in the cover soil.

Lastly, it should be noted that with large settlements exacerbated by the possibility of seismic activity, flexibility of final cover penetrations and appurtenances is advisable. Figure 1.6 shows the type of adapted "plumbing" that might be considered.

1.5.3 Inadequate Filtration

It is unfortunate that most writers on the subject of drainage layers neglect the importance of an accompanying filter layer. The authors are equally guilty, since the cross section on Figure 1.3 conveys the same assumption. Clearly, a protection soil cannot be placed directly on the drainage layer without a filter layer between the two materials, unless the drainage layer meets filter criteria with the adjacent soil. For natural soil drainage layers, the filter layer can either be a soil (typically a well graded sand) or a geotextile. For geosynthetic drains, the filter layer must be a geotextile since it has to act as a separator as well as a filter.

In either case, natural soil or geotextile, the filter must be properly designed. Filtration rules and guidelines are well advanced in both the geotechnical and geosynthetics literature, and well within the state-of-the-practice. Without proper design and subsequent proper installation, much less without any filter material whatsoever, the drainage layer is destined to become excessively clogged and the likelihood of instability heightened as discussed in the previous section.

1.5.4 Improper Gas Management

The degradation of the contents of municipal solid waste landfills produces a number of gases. For decomposing MSW landfills, methane and carbon dioxide are the principal gases. They are produced in roughly equal quantities, each comprising 40% to 50% of the decomposition gas. Those gases lighter than air, e.g., methane, rise in the waste only to impinge upon the underside of the barrier in the cover. It has been estimated (EPA 1981) that methane produced in a municipal solid waste landfill is generated over a long period of time. By that agency's estimates, 30% is produced in 2 years, 90% is produced in 80 years, and 99% is produced in 160 years. Some designers vent the cover system with vertical, perforated collection pipes (typically 5 pipes per 2 hectares) without a gas collection layer. This approach can only be justified if the waste itself is highly permeable to gas. In this case, a deep collection system using vents into the waste is possible. However, a gas collection layer, tapped by relatively shallow vent pipes, is the recommended approach for most situations.

Thus, it is usually necessary to provide a gas collection layer that is properly graded and that is in contact with vents penetrating the various materials in the cover, as shown in Figure 1.7a. The gas vents are either passive (vented to the atmosphere or flared) or active (collected with a vacuum pump and frequently used for energy production). Without such a venting system, the incidence of gas bubbles, as seen in Figure 1.7b, is to be expected. Note that with a geomembrane in the cover system, even nominal amounts of gas can be troublesome, causing uplift. Even if the geomembrane is not physically lifted, positive gas pressure beneath the geomembrane can lower the normal stress at the interface between the geomembrane and underlying material (e.g., GCL), thereby reducing interface shear strength and potentially contributing to a slope failure.

1.5.5 Long-Term Erosion

Water or air moving over the surface of the final cover can cause serious erosion involving soil particle transportation and downgradient deposition. Mitchell (1976) illustrates the phenomena in Figure 1.8. This sequence of events results in sheet, rill, and eventually gully erosion, which can have very serious consequences. Cases of excessive erosion of landfill covers have occurred. For example, at one site, more than a meter of cover soil was eroded (down to the underlying compacted clay liner) over several hectares within a four-month period. To prevent such incidents, the surface of the soil must be protected against water and/or wind erosion. This protection varies from local vegetation in humid areas, to rock armoring in arid areas. Several geosynthetic erosion control materials are available and can be used as temporary or quasi-permanent erosion control systems. Three categories for such materials are the following:

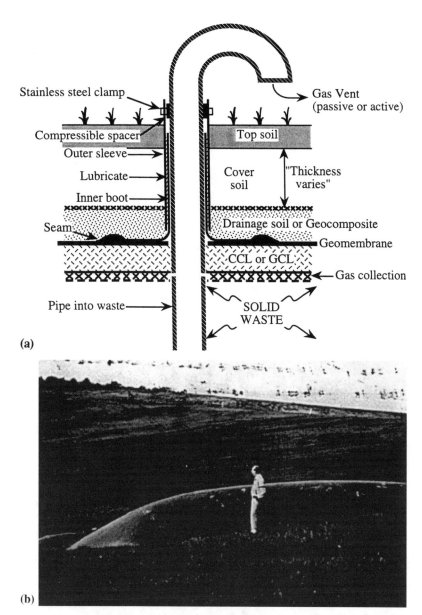

Stainless steel clamp

Gas Vent (passive or active)

Compressible spacer

Top soil

Outer sleeve

Lubricate

Cover soil

"Thickness varies"

Inner boot

Seam

Drainage soil or Geocomposite

Geomembrane

CCL or GCL

Gas collection

Pipe into waste

SOLID WASTE

(a)

(b)

FIG. 1.7. Gas Management Vent and Bubble at Municipal Solid Waste Landfill: (a) Gas Vent for Use in Passive or Active Systems Including Flexible Sleeve to Accommodate Waste Settlement; (b) Gas Bubble in Cover of Municipal Solid Waste Landfill

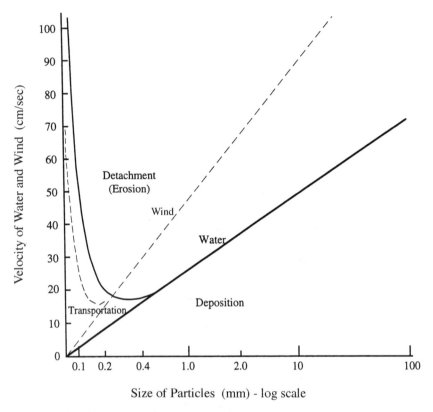

FIG. 1.8.   Relationships between Erosion Mechanisms (Water or Wind) and Respective Soil Particle Behavior, after Garrel (1951) (Extrapolated by Authors)

- temporary erosion and revegetative materials (TERMs),
- permanent erosion and revegetative materials (PERMs), which are soft-armor related, and
- permanent erosion and revegetative materials (PERMs), which are hard-armor related.

More information will be presented in Chapter 2.

1.5.6 Possible Final Use and Aesthetics

Closed landfills represent some of the largest human-made structures that exist. Furthermore, they are often located near populated areas where visibility by the public is high and potential utilization of the closed site would be an advantage. In this latter regard, it is important to recognize that, with a geomembrane in the final cover, the unpleasant odor of methane is essentially eliminated. Of course, a closed landfill represents a quasi-stable system, with on-going settlement, but uses can be found nevertheless.

Mackey (1996) presents case histories of closed landfills which are being successfully used for golf courses, sport fields, and walking/jogging/biking paths. Even further, one can envision highly visible final covers to be constructed as massive artworks. The New Jersey Meadowlands Commission has plans for sculpting one of its closed landfills into the shape illustrated in Figure 1.9. It is complete with flares and visual pathways, all with an astrological motif.

1.6 QUALITY CONTROL AND QUALITY ASSURANCE

Subsequent chapters of this book will confirm that multi-layer final covers of landfills are complex structures that are relatively difficult to construct. Although the construction industry is quick to respond to difficult projects, the technology is relatively young and construction quality control (CQC) is still emerging. Thus, it is necessary to have an additional level of monitoring of the construction activities as they take place. This is known as construction quality assurance (CQA). These two activities are defined in Daniel and Koerner (1995) as follows:

> • *Construction Quality Control (CQC):* A planned system of inspections that is used to directly monitor and control the quality of a construction project. Construction quality control is normally performed by the geosynthetics installer or, for natural soil materials, by the earthwork contractor, and is necessary to achieve quality in the constructed or installed system. Construction quality control (CQC) refers to measures taken by the installer or contractor to determine compliance with the requirements for materials and workmanship as stated in the plans and specifications for the project. Although there has been a long

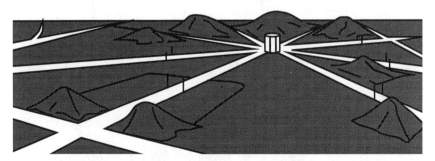

FIG. 1.9. *Artist Rendering of Topography of Final Cover [After Meadowlands Redevelopment Authority, Pinyon (1987)]*

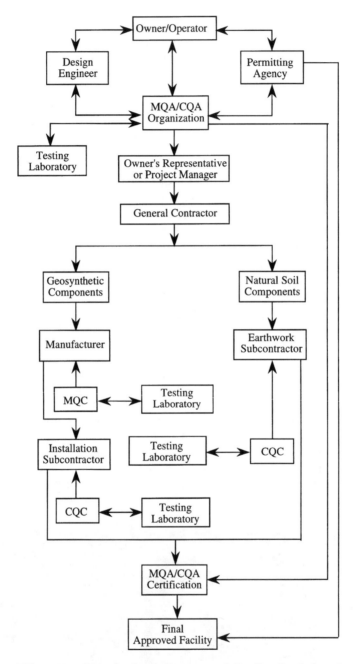

FIG. 1.10. *Organizational Structure of Quality Control and Quality Assurance Activities*

history of CQC in the geosynthetics industry, earthwork contractors have historically relied heavily upon information provided by the CQA process to control their operations, and have traditionally not had strong, internal CQC programs.

- *Construction Quality Assurance (CQA):* A planned system of activities that provides the owner and permitting agency assurance that the facility was constructed as specified in the design. Construction quality assurance includes inspections, verifications, audits, and evaluations of materials and workmanship necessary to determine and document the quality of the constructed facility. Construction quality assurance (CQA) refers to measures taken by the CQA organization to assess if the installer or contractor is in compliance with the plans and specifications for a project.

These two activities control the entire construction process and are paramount in achieving a final cover system as planned and designed. For the geosynthetic materials, there is an additional need for factory manufacturing quality control (MQC) and manufacturing quality assurance (MQA). The interaction of all of these activities in the context of an organizational structure for a typical project is illustrated in Figure 1.10. Additional detail will be presented in Chapter 7.

1.7 REFERENCES

Boschuk, J. J. (1991). "Landfill Covers: An Engineering Perspective," GFR, Vol. 9, No. 2, IFAI, March, pp. 23–34.

Corbet, S., et al. (1997). "Overview of USA and German Practices in Geosynthetic Aspects of Waste Containment Systems," Jour. Geotextiles and Geomembranes, (to appear).

Daniel, D. E., and Koerner, R. M. (1995). *Waste Containment Facilities: Guidance for Construction Quality Assurance and Quality Control of Liner and Cover Systems,* ASCE Press, New York, N.Y.

Edgers, L., Noble, J. J., and Williams, E. (1990). *A Biologic Model for Long Term Settlement in Landfills,* Tufts University, Medford, MA.

Galenti, V., Eith, A. E., Leonard, M. S. M., and Fenn, P. S. (1991). "An Assessment of Deep Dynamic Compaction as a Means to Increase Refuse Density for an Operating Municipal Waste Landfill," Proc. on the Planning and Engineering of Landfills, Midland Geotechnical Society, UK, July, pp. 183–193.

Garrels, R. M. (1951). *A Textbook of Geology,* Harper Brothers, New York, NY.

Gartung, E. (1996). "Landfill Liners and Covers," Proc. Geosynthetics: Applications, Design and Construction, A. A. Balkema, Maastricht, The Netherlands, pp. 55–70.

Gartung, E. (1995). "German Practice for Landfills," Proc. 1st Landfill Containment Seminar, Glasgow, Scotland, November 28, pp. 8–24.

König, D., Kockel, R., and Jessberger, H. L. (1996). "Zur Beurteilung der Standsicherhert und zur Prognose der Setzungen von Mischabfalldeponien," Proc. 12th Nürnberg Deponieseminar, Vol. 75, Eigenverlag LGA, Nürnberg, Germany, pp. 95–117.

Mackey, R. E. (1966). "Is Thailand Ready for Lined Landfills?" Geotechnical Fabrics Report, Vol. 14, No. 7, September, IFAI, pp. 20–25.

Mackey, R. E. (1996). "Three End-Uses for Closed Landfills and Their Impacts to the Geosynthetic Design," Proc. GRI-9 Conference on Geosynthetics in Infrastructure Enhancement and Remediation, R. M. Koerner and G. R. Koerner, Eds., GII Publ., Philadelphia, PA, pp. 226–244.

Pinyan, C. (1987). "Sky Mound to Raise from Dump," ENR, June 11, pp. 28–29.

Rumer, R., and Mitchell, J. K., Eds. (1996). Assessment of Barrier Containment Technologies, Publ. PB96-180583, National Technical Information Services (NTIS), Springfield, VA, 437 pgs.

Schroeder, P. R., Lloyd, C. M., and Zappi, P. A. (1994a). "The Hydrologic Evaluation of Landfill Performance (HELP) Model User's Guide for Version 3," U.S. Environmental Protection Agency, Office of Research and Development, Washington, DC, EPA/600/R-94/168a.

Schroeder, P. R., Dozier, T. S., Zappi, P. A., McEnroe, B. M., Sjostrom, J. W., and Peyton, R. L. (1994b). "The Hydrologic Evaluation of Landfill Performance (HELP) Model Engineering Documentation for Version 3," U.S. Environmental Protection Agency, Office of Research and Development, Washington, DC, EPA/600/R-94/168b, 116 pgs.

Spikula, D. (1996). "Subsidence Performance of Landfills: A 7-Year Review," Proc. GRI-10 Conference on Field Performance of Geosynthetics and Geosynthetic Related Systems, Geosynthetic Information Institute, Philadelphia, PA, pp. 237–244.

Stief, K. (1996). "Factors For and Against Final Covers in Landfills," Proc. VDI Seminar on Final Covers in Landfills, Karlsruhe, Germany, Oct. 9–10, pp. 1-1 to 1-18 (in German).

Stief, K. (1986). "The Multibarrier Concept in Landfill Construction," Müll und Abfall 1, pp. 15–20 (in German).

TA Abfall: Second General Administrative Provision to the Waste Avoidance and Waste Management Act, Part 1: Technical Instructions on the Storage, Chemical, Physical and Biological Treatment, Incineration and Disposal of Waste Requiring Particular Supervision of 12.3.1991, Müll-Hanbuch, E. Schmidt, Berlin, Vol. 1, 0670, pp. 1–136 (in German).

TA Siedlungssabfall: Third General Administrative Provision to the Waste Avoidance and Waste Management Act: Technical Instructions on Recycling, Treatment and Storage of Municipal Waste of 14.51993, Müll-Handbuch, E. Schmidt, Berlin, Vol. 1, 0675, pp. 1–52 (in German).

U.S. Environmental Protection Agency (EPA) (1981). Workshop Report on Landfill Gas Utilization, Los Angeles, CA.

U.S. EPA (1996). *Landfill Bioreactor Design and Operation*, EPA /600/ R-95/146, September, 230 pgs.

CHAPTER 2

INDIVIDUAL COMPONENTS OF FINAL COVER SYSTEMS

The six basic components of a final cover system were shown in Figure 1.3. The selection of materials and design of each of the individual components is discussed in this chapter. Examples of specific cover designs, i.e., the complete system, will be presented in Chapter 3.

2.1 SURFACE LAYER

The primary function of the surface layer is to provide separation between underlying components of the cap and the surface environment. Although vegetated soil is the most commonly used material for the surface layer, cobbles have been used at arid sites where it is difficult to maintain a vegetative cover. Other materials, both natural and synthetic, have occasionally been used.

2.1.1 Design Considerations

The designer has at least seven essential decisions to make in design of the surface layer. There may be others on a site-specific basis.

1. What materials will be used to construct the surface layer?
2. How thick will the surface layer be?
3. How will the layer be constructed?
4. For vegetated covers, which plants will be established?
5. Will a geosynthetic erosion layer be employed at the surface?
6. How will the surface layer be maintained?
7. If erosion occurs, will the rates be acceptable?

Some of the above design decisions are greatly impacted by the slope of the final cover and by the surface water control measures that are employed. It is considered good engineering practice to maintain slopes of 2 to 5% (after accounting for settlement) to promote runoff of surface water.

The definition of slope inclination is shown in Figure 2.1. As shown in Figure 2.2, many final covers have a relatively flat slope section on top and steeper slope sections on the side. The potential for erosion is greater on the relatively steep slopes.

It is necessary erosion control practice to provide for surface water diversion on long, uniform slopes that are steeper than 5%. Such diversion, illustrated in Figure 2.3, typically consists of benches perpendicular to the slope, which intercept the sheet flow along the slopes. As flow accumulates in the benches, it is often directed toward downchutes. Downchutes convey water to lower benches or away from the cover entirely, e.g., to streams, retention ponds, or sewers. The required spacing of benches and downchutes is controlled by the design storm and runoff coefficient. Typically, in areas that receive significant rainfall, benches are constructed approximately every 10 m of elevation change, or every 30 m measured along the slope. These distances, however, should be based on local factors and should not be arbitrarily established. Leaving out benches altogether on long slopes (lengths greater than approximately 30 to 50 m measured along the slope) with a topsoil surface layer is an invitation for excessive erosion conditions. Erosion rills forming gullies as deep as 1 m can develop, and hundreds of cubic meters of soil can be washed away in a few days.

2.1.1.1 Materials. While soil is by far the most common material for the surface layer, there are other possibilities in unique circumstances.

2.1.1.1.1 Topsoil. Soils are classified using either engineering or agricultural classification systems. The agricultural system, developed by the U.S. Department of Agriculture (USDA), is summarized in Figure 2.4. Soil is classified based on the relative amounts of sand, silt, and clay. Note that a mixture of sand, silt, and clay is called a "loam."

The preferred soil to use for support of local plants is locally available soil. In general, however, soils that promote and sustain plant growth are typically loamy soils (recall Figure 2.4). The sand in the loam provides a stable matrix that does not shrink and crack when desiccated and provides good drainage. Some fine material (silt and clay) is essential for moisture retention. Thus, a loamy soil is generally ideal.

It is also very helpful if the top soil contains adequate organic matter and plant nutrients. If not, supplements (fertilizers) may have to be added.

Sometimes the surface layer consists of two materials: top soil as one component and a different soil beneath the top soil. It is simpler, from both a design and construction perspective, to use a single material, but if top soil is scarce, economics may require minimizing its volume by incorporating a different soil beneath the topsoil layer.

Some covers are required to be light-weight to minimize settlement of underlying waste or soil. Light-weight agricultural and architectural soils

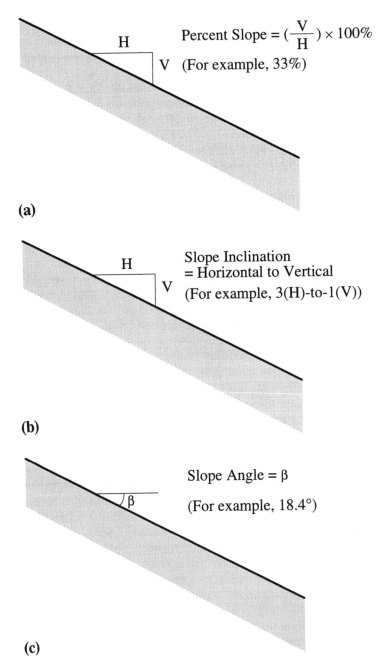

FIG. 2.1.   *Definitions Used to Describe Cover Soil Slopes*

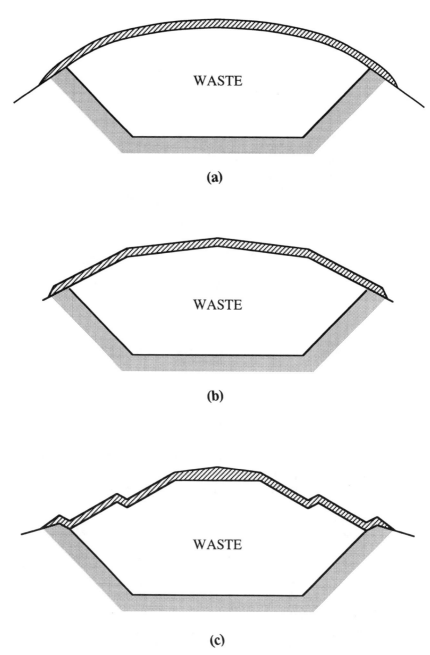

(a)

(b)

(c)

FIG. 2.2. Various Types of Final Cover Configurations

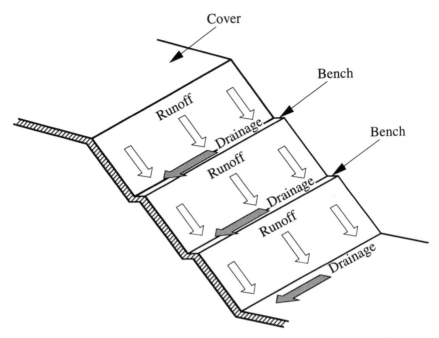

FIG. 2.3. *Isometric Sketch of Possible Runoff Pattern from Benched Constant Slope Angle Final Cover*

are available. These light-weight soils are manufactured by heating clay or shale in a kiln (to make so-called "light-weight" aggregate) and then screening the produced material to achieve the desired gradation.

Local agricultural specialists should be consulted in evaluating the soil to be used in the surface layer. The type of soil to be used may depend on the type of vegetation that will be planted.

Consideration should also be given to the tendency of the soil to be eroded. Erosion of the surface layer has been a common problem in final covers. Gullies extending to a depth of approximately 100 to 200 mm are not unusual. Swope (1975) studied 24 landfill covers in the United States and found that 33% had slight erosion, 40% moderate erosion, and more than 20% had severe erosion. Johnson and Urie (1985) report that erosion can be made more severe by the installation of a hydraulic barrier layer within landfill covers, which, without an overlying drainage layer, can cause the overlying soils to become soaked. This tends to increase runoff after heavy rains, especially if vehicles drive on the surface of the cover. Although erosion problems are often fairly minor and can usually be handled with routine maintenance, there have been instances of major erosion

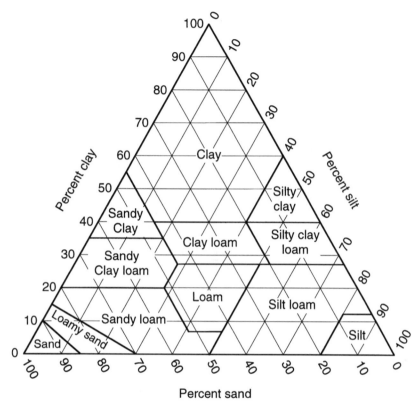

FIG. 2.4. *Soils Classification from Agricultural Perspective (after U.S. Department of Agriculture)*

that displaced hundreds of cubic meters of soil from inadequately protected landfill covers. Erosion potential must be given careful consideration.

The erosion potential of various soils was summarized in Figure 1.8. In general, the larger the size of individual soil particles, the less vulnerable the soil is to erosion. Clays, however, which have the smallest particle size, also tend to exhibit cohesion, i.e., they stick to each other, which helps to prevent erosion. Nevertheless, some sodium-rich clays do not adhere to one another very well and, therefore, are highly erodible. Such clay soils are called "dispersive clays." Silt has a small particle size but lacks cohesion. Silt is, therefore, almost always highly erodible. Dispersive clays and silts should not be used for the surface layer, unless it can be clearly demonstrated that erosion will not be a problem. The analysis of erosion rates is described later in this section.

2.1.1.1.2 Cobbles.   In arid locations, it may be very difficult to establish and to maintain consistent vegetation on the surface layer. Soil, if used without a sufficiently vibrant stand of vegetation, is vulnerable to excessive erosion from wind, water, or both. The solution in arid areas may be to armor the surface with gravel or cobbles. This type of armoring is called "rip-rap."

The material to be used for rip-rap is heavily dependent upon what is locally available. In earthwork projects, rip-rap is often the most expensive material used on the project. This is because sources of clean gravel or cobbles are fairly rare, which means that the stones must often be quarried from rock. Thus, before rip-rap is selected as the surface layer, the cost of the material should be established.

A conservative design for the average weight of rip-rap stone on slopes adjoining bodies of water is given by Hudson (1959).

$$W = \frac{\gamma_s H^3 \tan \alpha}{\lambda_D (G_s - 1)^3} \tag{2.1}$$

where

W = average weight of stone,
H = wave height,
α = slope angle,
λ_D = 2.73 (typical),
G_s = specific gravity of stone, and
γ_s = unit weight of stone.

Furthermore,

$$D = \sqrt[3]{0.699W} \tag{2.2}$$

where

D = average diameter (in m), and
W = stone mass in tonnes (1 tonne = 1000 kg).

Equation 2.1 is applicable to the design of slopes next to bodies of water, where waves challenge the rip-rap. Final covers rarely have waves impinging on them, but the experience with Equation 2.1 is so extensive that it should be considered. For instance, for a 3(H) to 1(V) slope, if the wave height is 1 m and G_s = 2.7, the required average mass of the stones is about 2 kg. Alternatively, Figure 1.8 may be consulted. For instance, if the water velocity is 0.5 m/s, the minimum particle size to prevent water erosion is about 10 mm. An empirical approach may also be useful if local experience with rip-rap on comparable slopes is available.

One of the problems with armoring the surface layer with rip-rap is an unfavorable impact on the water balance of the final cover system. The armor allows for only one-way moisture movement, i.e., precipitation percolates downward through the rip-rap, but evaporation is not encouraged because plants are not present to remove water from the subsoil and transpire it back to the atmosphere. The tendency of granular material to behave in this manner is utilized by gardeners who apply mulch to bare soil. The mulch allows water to percolate down to the underlying soil but shields the soil from evaporative loss of water.

Gravel surface covers are known to permit accumulation of water beneath the gravel (Kemper et al. 1994) and to promote deep drainage (Gee et al. 1992). Admixtures of fine and coarse materials, such as the silt-gravel admix developed at Hanford, Washington (Ligotke and Klopfer 1990), can provide a means of controlling erosion while maintaining a favorable water balance. With the silt-gravel admix, wind erosion removes a thin layer of silt at the surface, but the gravel particles are too large to be eroded by wind. Eventually, enough silt is eroded to leave a thin layer of gravel at the surface (a so-called "desert pavement"), with the undisturbed silt-gravel admix located no more than a centimeter or two from the surface. The silt sustains the growth of vegetation (the gravel does not interfere with plant growth), and the thin gravel layer that develops naturally at the surface stops wind erosion. The silt-gravel admix has been used in the final cover of a radioactively contaminated landfill site (Wing and Gee 1994), but only in relatively flat areas where water erosion was not a concern. Rock rip-rap was used on the steeper side slopes.

2.1.1.1.3 Asphaltic Concrete. Asphalt can be very permeable unless special attention is given to eliminating the air voids during mixing and application (Repa et al. 1987). To achieve low permeability, more asphalt is used (1.5 to 2.0 times the normal amount) than for roadway pavements. The authors are aware of three cover systems in which asphaltic concrete was used at the surface layer. One case was a cover used to close a radioactively contaminated site. The final cover system included a composite geomembrane/clay hydraulic barrier layer, a sand drainage layer, and a soil protection layer. The 1-ha site was located next to an office building on flat ground, and it was decided to pave the cover and to use it as a parking lot. A second use was in a landfill cover where a paved area was required for maintenance vehicles. The third project was a Superfund remediation project where there was unusual concern for minimizing or eliminating erosion. In the latter two cases, the asphalt was a "hydraulic asphalt" that contained a higher-than-usual amount of asphalt to reduce the hydraulic conductivity to below 1×10^{-7} cm/s.

However, the authors do not recommend a permanent, low-permeability asphaltic concrete barrier at the surface. The asphalt will degrade as a result of exposure to ultraviolet (UV) radiation and oxygen. If asphalt is used as a hydraulic barrier, it should be buried beneath a protection layer, not exposed at the surface; or if placed at the surface, only considered as a temporary hydraulic barrier layer.

2.1.1.1.4 Other Materials. Practically any material, including some construction demolition materials, could potentially be used as a material in a surface layer. However, if something other than soil or rip-rap is considered, it will generally be because there is a special desire or incentive for using a particular material, and these special circumstances will be known to the designer. Alternative materials should be considered if they are safe, stable, and can meet performance objectives of the final cover.

2.1.1.2 Thickness. The minimum thickness of the surface layer is established by typical construction tolerances. With earth moving equipment, it is usually not practical to attempt to construct a layer thinner than about 150 mm. If topsoil is used, the topsoil must also be thick enough to accommodate a healthy growth of plant roots. For shallow-rooted plants such as grass, a 150-mm-thick layer of topsoil usually provides adequate rooting depth. Thus, topsoil is almost always a minimum of 150 mm thick.

The design thickness of the surface layer may be greater than 150 mm, depending on many factors. If plants with roots that penetrate deeper than 150 mm will or could become established on the cover, a greater thickness of soil to accommodate root growth is recommended. The underlying protection layer (if present) may also accommodate plant roots, in which case 150 mm may be all that is needed for the surface layer. In many instances, the surface layer and protection layer are one type of material, making it impossible to distinguish one layer from the other. In such cases, the combined layers may be referred to as "cover soil" or "cover material." If the surface and protection layers are combined into a cover soil, then the minimum thickness of the cover soil could be controlled by required rooting depth. A typical thickness of the cover soil is 450 to 600 mm.

If rip-rap or other granular material is used for the surface layer, the thickness would usually be a minimum of 150 mm or the maximum particle size of the rip-rap, whichever is larger. If asphaltic concrete is used for the surface layer, the minimum thickness would be determined from an analysis of vehicular loading but would typically be in the range of 75 to 150 mm.

2.1.1.3 Construction. Construction of the surface layer is straightforward. If topsoil is used, the soil is only nominally compacted, if compacted at all, to maximize the ease with which plants can grow. Rock rip-rap is normally placed loose with little or no compaction. A gravel-soil admixture will require some compaction, but heavy compaction is neither necessary nor desired. If asphaltic concrete is used as the surface layer, normal road-building equipment is used.

2.1.1.4 Establishing Vegetation. Selection of plant species is an important consideration in the establishment of vegetation when it is selected as the upper component of the surface layer. The use of shrubs and trees is usually inappropriate because the root systems extend to a depth that would normally invade the drainage layer, or the barrier layer if it is low-permeability soil. Trees can also create problems if they are blown over, uprooting large masses of soil.

Most cover systems are seeded with grasses. Suitable plant species, such as grasses and low-growing plants, are available for various climates. The timing of seeding is also important to successful establishment of vegetation. Several references outline selection criteria (Lee et al. 1984; Thornburg 1979; and Wright 1976). These references provide essential information about plant species, seeding rate, time of seeding, and areas of adaptation.

The vegetated surface layer should generally meet the following specifications:

- Consist of locally adapted perennial plants;
- Be resistant to drought and temperature extremes;
- Contain roots that will not disrupt the drainage or barrier layers;
- Be capable of thriving in low-nutrient soil with minimum nutrient addition;
- Develop sufficient plant density to minimize cover soil erosion; and
- Be capable of surviving and functioning with little or no maintenance.

In landfill situations where the environment or other considerations make it inappropriate to maintain sufficiently dense vegetation, armoring material may be necessary for the surface layer. It is recommended that the armor material be capable of:

- Remaining in place and minimizing erosion of itself and the underlying soil component during extreme weather events of rainfall and/or wind;
- Accommodating settlement of the underlying material without compromising the purpose of the component; and
- Controlling the rate of soil erosion from the cover to acceptable levels (typically no more than 4.5 Mg/ha/yr (2 tons/acre/yr), as calculated by using the Universal Soil Loss Equation discussed later).

The thickness of the topsoil layer should depend on many site-specific factors, including climate, plant species to be grown on the cover, and soil materials to be used. As a minimum, the surface layer should be at least 150 mm thick, because this is typically the minimum depth of root penetration of most plant species. Additional recommendations include:

- A medium textured soil (e.g., loam) to facilitate seed germination and plant root development;
- Minimal compaction of the soil, to facilitate root development; and
- A final top slope, after allowance for settling and subsidence, of at least 3%, but no greater than about 5%, to facilitate runoff while minimizing erosion. Slopes steeper than 5% will generally require some type of erosion control material.

Local soils that are known to make a good growth medium are normally selected. Care should be taken not to select a soil that is too coarse grained (e.g., a clean sand) because the soil may retain too little moisture during drought periods, compromising the sustainability of the surface vegetation.

Numerous materials are available for use in the topsoil layer, including geosynthetic and natural erosion control materials, and blends of topsoil with additives, including sewage waste sludges. The use of alternative materials is encouraged where it can be demonstrated that such materials will perform in an equivalent or superior manner to more conventional materials.

2.1.2 Erosion Considerations

Excessive erosion has been a significant problem at many landfill final covers. The design should address the potential for short-term erosion, e.g., before a good stand of vegetation is established, and make use of temporary erosion-control measures as necessary. Erosion can be damaging not only to the final cover system but also to areas into which eroded soil is deposited. The timing for completion of construction of the final cover system with respect to the growing season for the grasses that will be planted on the final cover is very important. It may be impossible to initiate growth of the vegetative cover during certain parts of the year. In many cases, construction is completed at the end of the growing season, and many months pass before significant growth of vegetation can be established. In such cases, geosynthetic or natural erosion control materials are recommended. Unfortunately, there are situations where the geosynthetic erosion control materials were not included in the budget, and were therefore omitted, leading to development of significant erosion problems that required repair. Failure to consider temporary erosion control is a common oversight.

Silt fences are often placed at the toe of landfill cover slopes to minimize impact of sediments on roads or streams. A silt fence is intended to minimize the impact of erosion upon adjacent areas and is not by itself an appropriate preventive measure to stop erosion. Also, for cover systems at abandoned dumps, and even some new landfills, gas collection pipes may be laid on the surface of the cap to permit the collection of gases from gas vents. The pipes can channel surface runoff into isolated areas, where rill and gully erosion can be initiated. Additional erosion control measures, e.g., geosynthetic or natural erosion control materials, may be needed in such areas.

Long-term erosion should be an important consideration in the design of the surface layer. Although it is appropriate to select allowable rates of soil erosion on a project-specific basis, most designers follow the general guideline that erosion rates not exceed 5.5 MT/ha/year. To control short- and long-term erosion, benches are usually needed in the final slope (recall Figure 2.3) with surface water collection and diversion from the benches. Appropriate erosion control should be maintained on the benches, with surface water channeled downslope through downchutes that are designed to convey surface runoff with minimal erosion. The distance between benches has not been standardized because precipitation patterns vary enormously from one region to another. However, in areas when erosion has been routinely experienced in unprotected final covers or slopes, benches approximately every 10 m vertically or 30 m along the slope are a common design practice.

The most often used model for soil loss by erosion is the Universal Soil Loss Equation (USLE). The equation is given by

$$E = RK(LS)CP \qquad (2.3)$$

where

E = soil loss (tons per acre per year or tons per square kilometer per year, depending on constants used),
R = rainfall factor (dimensionless),
K = soil erodibility factor (dimensionless),
LS = length of slope or gradient factor (dimensionless),
C = vegetative cover factor (dimensionless), and
P = conservation practice factor (dimensionless).

Charts and tables describe the various factors involved (see Wischmeier and Smith 1960).

There are many limitations of the preceding equation. For example, it is not applicable to predict erosion from rill or gully-type runoff, small-localized sites, steep slopes, seasonal variations, or short-term water surges. While the equation is useful in a global sense, e.g., to predict large-scale farmland soil loss, it is generally felt not to provide meaningful guidance in the quantification of site-specific soil erosion. For this reason there are numerous ongoing activities to modify the equation for point-source soil loss by erosion.

In sloping terrain, diversion structures should be installed, if needed, to prevent overland flow of surface water onto the final cover of the facility.

To prevent ponding of rainwater, the final surface should be uniformly graded and sloped at least 3%, after allowance for settlement and subsidence. Slopes greater than 5%, however, are likely to promote erosion unless controls are included in the design.

2.1.3 Vegetation Details

Selection of plant species is an important consideration in the establishment of vegetation when it is chosen as the upper component of the top layer. The use of shrubs and trees is usually inappropriate because the root systems extend to a depth that would normally invade the drainage layer or the low-permeability layer. Trees can also create problems if they are blown over, uprooting a large mass of soil. Suitable plant species such as grasses and low-growing plants should be selected for local conditions. The timing of seeding is also important to successful establishment of vegetation. Several references outline selection criteria (Lee et al. 1984; Thornburg 1979; and Wright 1976). These references provide essential information about plant species, seeding rate, time of seeding, and areas of adaptation.

2.1.4 Surface Armor

In areas where vegetation is inappropriate or difficult to establish and maintain, other materials may be selected as the upper component of the top layer. The material should be selected with the objective of preventing deterioration of the cap due to wind, heavy rain, or temperature extremes, without reliance upon vegetation. The materials should be selected to prevent erosion of the cover and yet allow for surface drainage.

Several materials have been suggested and in some cases used in lieu of vegetation, including rock rip-rap, cobbles, and concrete blocks of various designs. Asphalt, concrete, or special erosion control materials such as those filled with concrete might be used if promoting runoff is a prime objective, but they are likely to deteriorate, for example, by cracking due to thermal effects and subsidence deformation, thus causing concern for their long-term performance.

2.1.5 Maintenance

The maintenance of the surface of final covers is eminently straightforward, yet difficult to do consistently after closure, and particularly after the end of the post-closure care period.

In arid areas, without vegetation, the situation is one of visual observation for signs of erosion. Rill and gully erosion are readily observable, yet wind erosion is often of the sheet variety and more subtle to observe. Perhaps downwind deposition of soil is the best indicator.

In humid areas, with vegetation over the final cover, the situation is more difficult to assess. At the minimum, vegetation height should be controlled so that rill and gully erosion is observed and corrected. This requires grass and weed cutting on a regular basis. Trees should be cut while still small, so that root systems do not mature and do not challenge the drainage system beneath the protective layer.

If the final cover is being used in a productive manner, the maintenance issue can be folded into the land use plan. If not, which is generally the case, constant vigilance is necessary and action must be taken when justified.

2.2 PROTECTION LAYER

The protection layer lies directly beneath the surface layer, and in some cases can be combined with it. In the following text, it is described as a separate and distinct layer.

2.2.1 General Considerations

The protection layer underlies the surface layer and serves both to protect the underlying components and to store water that has percolated through the surface layer. While the large majority of covers have a protection layer, some do not. In such cases the surface layer may be placed directly on a drainage layer or barrier layer. In most cases, this design approach is not appropriate because erosion gullies will cut through the surface layer (if it is relatively thin) and expose the underlying layer. Many materials (e.g., geosynthetics) will be damaged by prolonged exposure. If minimum percolation through a final cover system is required, e.g., for hazardous waste landfills, then a protection layer is usually required. If the cover design calls for a vegetated surface layer, the protection layer should be capable of indefinitely sustaining growth of plant species that will minimize erosion.

The required thickness of the protection layer depends on what is to be protected and how much protection is needed. If underlying vegetated surface layers, the protection layer should have adequate thickness to:

- Accommodate the root systems of plants that are expected to grow on the surface layer;
- Minimize intrusion of plant roots into underlying components of the cover system;
- For most locales, provide adequate water-holding capacity to attenuate rainfall infiltration to the drainage layer and to sustain vegetation through dry periods;

- Prevent damage to underlying components of the final cover system from borrowing animals;
- Protect, as necessary, the underlying components from the potentially damaging effects of moisture (wet-dry) and freeze-thaw cycling; and
- Provide sufficient soil thickness to allow for expected long-term erosion loses.

Plant roots or burrowing animals (collectively described as biointruders) may disrupt the integrity of the drainage and barrier layers. The drainage layer may be especially susceptible to the intrusion of plant roots, which could interfere with the drainage capability of the material. The danger of geomembrane penetration by plant intrusion has not been proven and is thought to be highly unlikely for most geomembrane materials. Burrowing animals may be a greater threat to geomembranes, if a threat indeed exists. In the absence of a geomembrane, the low-permeability soil layer (GCL or CCL) could be exposed to both root and animal penetration.

2.2.2 Materials

A variety of soil types are used for the protection layer, depending on locally available materials. Medium-textured soils, such as loam, have the best overall characteristics for seed germination and plant root system development. Fine-textured soils, such as clays, are often fertile but may be beset by management problems such as puddling of water on the surface or difficulty in initial establishment of plant cover during wet periods. Sandy soils may be a problem due to low water retention and loss of nutrients by leaching. It may be cost-effective to stockpile the topsoil initially removed from a borrow site for later use as surface layer material.

Other materials, e.g., cobbles, may be used in the protection layer if needed for special applications, as discussed in the next subsection.

2.2.3 Thickness

The minimum required thickness of the protection layer depends on many site-specific factors, including:

- Need to support growth of vegetation;
- Maximum depth of frost penetration;
- Need to store water in the protection layer;
- Need to prevent accidental human intrusion, penetration by burrowing animals, or root penetration into underlying materials;
- Need to protect underlying layers from desiccation; and

· Need to provide other types of protection unique to a particular waste (e.g., attenuate radiation if the waste is radioactive).

2.2.3.1 Support Growth of Vegetation. In order to support growth of vegetation, the soil must retain adequate moisture, even during periods of drought. Use of sandy soils for the protection layer material can be dangerous—sandy soils can dry out and may not contain sufficient moisture to support growth of plants. There have been situations reported in which the surface soils have dried to the extent that maintenance personnel were forced to irrigate the surface of the cover in order to maintain adequate soil moisture for growth of grass. (Post closure maintenance of a vegetative cover was required in the permits for the facilities.) The addition of water to the surface of a cover is not recommended if it can be avoided because one purpose of the cover is to stop or severely restrict the infiltration of water into the underlying waste. The use of a drainage layer beneath the protection layer can actually add to this potential problem by helping to drain and dry the protection layer. The solution to the problem is to use soils in the protection layer with adequate silt and clay content to retain sufficient moisture for plant growth, even during drought periods. Loams are recommended.

2.2.3.2 Frost Penetration. The protection layer is often designed with the intent of preventing underlying layers from freezing. As will be discussed, freeze-thaw cycles damage most compacted clay materials. Thus, a compacted clay liner is usually placed below the depth of maximum frost penetration. It may be prudent to prevent the drainage layer (if one is present) from freezing as well. The primary use of an underlying drainage layer in many final cover systems is to dissipate pore water pressures that tend to promote slope instability. By draining water from the slope in the drainage layer, the stability of the slope is enhanced. The stabilizing impact of drainage in slopes can be dramatic, causing a major difference in the factor of safety (i.e., doubling or halving the value depending on whether drainage is or is not controlled). If the drainage layer freezes, its function is destroyed for part of the year. During the thaw period, it is particularly important that the drainage layer function properly and that the protection layer be sufficiently thick to provide the protection that is required.

There are several techniques available for estimating depth of frost penetration. One technique is to use standard frost penetration maps (see Figure 2.5). Local experience is sometimes used, as are computer simulations.

2.2.3.3 Storage of Water. Most of the meteoric water that contacts the surface of a final cover system infiltrates into the underlying soil, where the water is retained by capillary forces and stored. The ultimate fate of

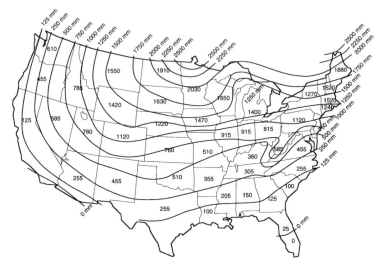

FIG. 2.5. Contours of Maximum Frost Penetration Depths (in mm) in Continental U.S. and State Averages (in mm) (Redrawn from U.S. Department of Commerce Weather Bureau Data)

most of this water is evapotranspiration. It is critical that the near-surface soils in a final cover system be capable of retaining significant moisture so that this moisture is stored and later (when weather and/or plant conditions are favorable) returned to the atmosphere via evapotranspiration.

The greater the percentage of fines in the soil, the greater is the moisture retention capacity. Clay soils, however, are generally not used for the protection layer because these soils tend to be more difficult to work with than sandier materials and because clays can dry and crack, providing an infiltration pathway that enables water to bypass the clay matrix and thereby bypass storage. However, some regions, such as the Texas Gulf coast, contain almost nothing but highly plastic clays near the surface. In such cases, there may be no practical alternative to the use of a heavy clay soil. If available, a loamy soil is the best for good moisture retention but good workability and resistance to desiccation cracking.

2.2.3.4 Accidental Human Intrusion. Accidental human intrusion has generally not been a design consideration for municipal solid waste, hazardous waste, or abandoned dumps requiring correction action. Essentially, the only type of waste for which accidental human intrusion has been a design factor is radioactive waste. It is not clear why radioactive waste has been singled out as the only type of waste for which accidental human intrusion is a design consideration. Human intrusion into municipal solid waste or hazardous waste could be extremely dangerous (potentially fatal) to the intruder. Perhaps the concern about radioactive waste is that symptoms of health impact might not appear immediately for a human intruder—health impacts from exposure to radiation are often chronic, may not appear for many years, and may have a bioaccumulation factor in certain types of human or animal tissue, e.g., accumulation in bones. Also, rightfully or not, humans tend to be more frightened of radioactive waste than other types of waste.

When human intrusion has been considered, the principal concern is accidental, e.g., routine excavation to lay a buried pipeline or to excavate a basement for a home. The cover can be thickened to approximately 5 m or more to account for these types of routine excavations.

2.2.3.5 Burrowing Animals.

Owing to the heightened concern over burrowing animals in radioactive waste, some thought has been given to possible scenarios. For example, Sutter et al. (1993) summarize the effects that burrowing animals can have on final covers as follows:

- They may burrow through the cover, resulting in direct channels for movement of water, vapors, roots, and other animals.
- Even when they do not penetrate the waste, burrows may increase the porosity of the soil thereby increasing rates of infiltration.
- Conversely, animals construct their burrows for natural ventilation, which may dry the soil and decrease water intrusion.
- Animals may become externally contaminated or consume the waste thereby spreading the waste in their feces, urine, and flesh, and increasing decomposition of the waste.
- They may carry waste directly to the surface during excavation.
- By working the soil and transporting seeds, they may hasten establishment of deep-rooted plants on the cover.
- They cast soil on the surface thereby increasing erosion of the cover.

Research by Cline (1979) and Hokanson (1986) found that if objects, such as cobbles, placed in a burrowing animal's path are sufficiently large and/or tightly packed, the animal's progress is effectively stopped.

Physical barriers to stop burrowing animals have been considered for closure of areas contaminated with radioactive materials because there have been reported instances of the spread of radioactivity at the surface as a result of animal burrows into the waste. Unfortunately, there are no full-scale applications available that prove the effectiveness of a biotic barrier in a landfill situation or even help to establish design guidance. Therefore, the design of such a barrier must rely on the results of small-scale field experiments. Experiments with barrier layers of cobbles have been carried out in arid or semi-arid situations with plants unique to such habitats (Cline 1979 and Cline et al. 1982). Some research suggests that 900 mm of cobbles, or 150 mm of gravel over 750 mm of cobbles, may be effective in stopping root penetration of some deep-rooted plants (DePoorter 1982). Such materials and thicknesses may also be effective in stopping the invasion of burrowing animals. The biotic barrier layer directly underlies the soil component of the surface layer, perhaps separated by a geotextile serving as a combined filter and separator.

2.2.3.6 Root Penetration.   Different plant species develop root systems that penetrate to highly variable depths. Root systems of shallow-rooted grasses may penetrate no deeper than 100 to 150 mm into the subsoil. Grasses with deeper root systems may have roots that penetrate to depths of 300 to 500 mm. Root systems of shrubs can penetrate to depths of greater than 1 m. Roots of some desert species can penetrate many meters into the subsurface. Trees also have root systems that can penetrate many meters into the subsurface, and this should be prevented via routine maintenance.

Climate can also have a major impact on the depth of root penetration, and even the materials into which roots penetrate. Roots generally seek out soils that contain moisture. Roots will not, as a general rule, penetrate into dry soils. In soil profiles containing a fine-textured soil overlying a coarse textured soil (e.g., gravel), roots will remain in the relatively moist, fine-textured soil and will not penetrate into the coarse-textured soil as long as it is dry. If the coarse-textured soil becomes soaked, then the roots will seek moisture in the coarse-textured soil.

If vegetation is planted on the surface layer, the combined thickness of the surface and protection layer must be large enough to support growth of plants. In theory, a total combined thickness of 150 mm may be sufficient for shallow-rooted grasses. However, over time, deeper-rooted grasses will often inevitably become established, rendering the soil too thin for adequate support of vegetation. In general, a minimum combined thickness of the surface layer and protection layer is 450 to 600 mm for most situations. However, the thickness should be determined based on local soils, plants, and climate.

Sutter et al. (1993) summarize the damage that plant roots can have as follows:

- They may penetrate the barrier layer.
- Decomposing roots leave channels for movement of water and vapors.
- Roots may dry clay layers, causing shrinking and cracking.

- Uprooted trees may leave depressions in the cap.
- Roots may enter the wastes, take up constituent chemicals, and transport them to above ground components.
- Roots may modify the waste by increasing decomposition rates and by releasing chemicals that mobilize metals.
- Plants may decrease leaching by taking up water.

Sutter et al. (1993) also provide examples of many of these potential problems.

Hokanson (1986) found that large void spaces, which lack water and nutrients, within a layer of stone reduced the intrusion of plant roots. On the other hand, a layer of very coarse materials, at least in arid areas, may favor the growth of grasses by impeding the downward percolation of moisture, thus helping to retain it in the top soil layer.

Cline et al. (1982) also examined the effectiveness of several phytotoxins impregnated into or onto geotextiles which were placed within the protection soil layer. Some of them met the goal of being effective in stopping the downward progress of root growth, with no other effects. Some of the phytotoxins killed the plants when the roots encountered the fabric, while others had no effect. Obviously, a chemical biotic barrier must be chosen carefully, if at all, to avoid potentially adverse environmental effects.

A polymeric herbicide carrier/delivery system, used to release herbicide, as discussed by Cline et al. (1982), might be installed within a protection layer just above the drainage layer to stop the intrusion of roots below the system. The system would contain a herbicide designed to be released slowly over many years. Note, however, that a regulatory agency would probably be reluctant in approving this alternative because it may introduce a hazardous waste to the cover system and/or it may not last through the post-closure period.

A layer of dry, granular material (e.g., gravel or cobbles) will serve to impede the downward movement of plant roots, particularly if the overlying layer is composed of a finer-textured soil that tends to retain more water than the granular material. In arid climates, this approach can be very effective in causing plant roots to remain in the moisture-retaining, finer-textured soil.

2.2.3.7 Desiccation of Underlying Layers.

The protection layer may be designed to protect underlying layers from desiccation. However, if no geomembrane is used in the hydraulic barrier layer, the problem of protecting underlying layers from desiccation by a suitably thick protection layer is particularly difficult. Geomembranes provide extremely effective barriers to water and vapor movement and, in conjunction with cover soil over the geomembrane, provide by far the most effective means of protecting underlying soils from desiccation. Although other barrier materials such as sprayed-on asphalt could also be used, geomembranes have thus far been used almost exclusively for thin, flexible barriers to liquid and gas migration.

Experience has shown that severe desiccation can extend to depths of up to 1 m, and probably deeper. The only information that is available on field patterns of desiccation covers periods of observation of approximately 5 years. Over longer periods, the problem of protecting underlying soils from desiccation would be even more difficult because the probability of an extreme weather event occurring increases with the passage of time. In reality, we do not know how deep a protection layer needs to be to prevent desiccation of underlying layers (particularly a compacted clay liner) if a geomembrane or similar vapor barrier is not used. Further, the required depth would no doubt vary considerably from one site to another. Because of the lack of information, a conservative approach is recommended. In most situations, it appears that the protection layer would have to be at least 2 m thick, and perhaps as much as 5 m thick, to protect underlying layers from seasonal desiccation in the absence of a geomembrane. But it is emphasized that if it is desired to provide such protection, a geomembrane in the underlying barrier layer combined with adequate cover soil (see Chapter 3) is the best way to minimize the possibility of desiccation of underlying materials.

2.2.3.8 Radiation Protection.

Some radioactive wastes emit radon in the form of a gas. To minimize release of radon to the environment, the cap may be designed as a barrier to radon. Because radon is in gaseous form, the design objective is to limit release of gas to the surface. Thus, a gas barrier is needed.

A geomembrane can provide an excellent barrier to gas release, but many designers have been reluctant to include a geomembrane in a cap for radioactive waste for fear that the service life of the geomembrane will be inadequate. In the authors' opinion, this is a mistake. Although the geomembrane will not last forever, a properly selected and appropriate geomembrane formulation can last for centuries, and because the cost of geomembranes is very low, a geomembrane can provide an extremely cost effective means of control of radon gas.

Instead of relying on geomembranes, designers have placed more weight on natural soils, since they theoretically have an indefinite service life. However, for a soil to be a barrier to gas diffusion, the air-filled voids have to be discontinuous (called "occluded" soil air). Gas diffuses extremely slowly through wet soil, and the design objective is to employ clay-rich soils in the cover system as a radon barrier. The radon barrier could be in the protection layer, or it could be in the hydraulic/gas barrier (or even in both). The best diffusion barriers tend to be relatively thick, and radon barriers need to be at least several meters thick. The real problem with clay materials, though, is that they must be wet and free of cracks to work. Over a design life of hundreds of years, the reality is that engineers cannot be sure that the clay will always be in this condition.

2.2.4 Capillary Barriers

A fairly recent development is the use of a layer of fine-textured soil overlying a layer of coarse-textured soil to form a "capillary barrier." The idea is as follows. Soil moisture in the subsurface reaches equilibrium when the soil water potential is the same throughout. If a layer of fine- and coarse-textured soil are in equilibrium and there is no movement of water between the layers, the two layers will have the same soil water potential. For a given soil water potential, a coarse-grained soil will tend to have a much lower water content, i.e., be much drier, than a fine-textured soil. The hydraulic conductivity of unsaturated soils decreases exponentially with decreasing water content because flow paths through thin films of water coating the soil particles in a dry soil are extremely tortuous. A dry gravel is actually much less permeable to water than a moist sand.

Thus, if the subsoils remain unsaturated, a fine-textured soil overlying a coarse-textured soil will tend to function with the uppermost soil layer retaining nearly all the soil moisture and the underlying layer serving as a barrier to water percolation due to its dryness. The contact surface between the fine- and coarse-textured soils can be sloped, much like at the interface between a drainage layer and underlying barrier layer. However, in a capillary barrier, lateral movement of water in the fine-textured soil occurs in an unsaturated state. The layer is sometimes called a "wicking layer." The concept has been shown to work in pilot-scale experiments (Nyhan et al. 1990; and Fayer et al. 1992). There are two concerns, however, with capillary barriers. One is that the fine-textured soil must not be allowed to migrate over time into the underlying coarse-grained soil. A geotextile, used as a separator, should be considered for placement beneath the fine-textured soil and above the coarse-textured soil. For extremely long service lifetimes, fiber glass geotextiles have been considered for this application. The second concern is over periods of extremely high (relatively speaking) precipitation. In such a case, the capillary barrier concept may cease to function, at least temporarily, as the coarse-textured soil becomes moist and loses its water-impeding capability.

2.3 DRAINAGE LAYER

Water that penetrates through the cover soil may be removed from the cover system using a drainage layer.

2.3.1 General Considerations

As discussed in Chapter 1, a drainage layer serves three principal functions:

- Reduces the head of liquid on the underlying barrier layer, thereby minimizing the amount of water percolation into underlying layers, waste, or contaminated soil;
- Drains water from the overlying soil, allowing it to absorb and retain additional water; and
- Eliminates pore water pressures at the interface to the underlying barrier layer.

Perhaps the most serious problem to be avoided with drainage layers is excessive long-term clogging. This is the case for natural soil drains as well as geosynthetic drains. If, for example, the stability of the slope depends on proper functioning the drainage layer and the layer excessively clogs, a slope failure becomes a very real possibility. The drainage layer should be designed, constructed, and operated to function without excessive clogging over the lifetime of the facility. Several failures of covers are described by Boschuk (1991). Most of these failures were attributed to fundamental errors made in assessing the stability of the cover, or in several instances, lack of an analysis of the stability of the cover, particularly with respect to seepage forces.

Excessive clogging can be prevented by incorporating a filter layer of soil or geotextile between the drainage layer and the overlying soil/protection layer. The prevention of biological clogging by plant roots is usually accomplished by using suitably thick surface and protection layers, and by limiting vegetation to shallow-rooted species.

In arid locations, the need for, and design of, a drainage layer should be based on consideration of precipitation event frequency and intensity, and sorptive capacity of other soil layers in the cover system. It may be possible to construct a surface layer and protection layer that will absorb most, if not all, of the precipitation that infiltrates into that layer, eliminating the need for a drainage layer.

2.3.2 Materials

The materials for use in the drainage layer are either cohesionless soils or drainage geosynthetics.

2.3.2.1 Granular Materials. Granular materials used for drainage layers are almost always sand or gravel. Often, gravel forms the drainage material and sand is the filter material. Sometimes the drainage material inherently provides adequate protection against migration of soil particles from the adjacent soil and a separate filter is not required. However, filter criteria (described in the next subsection) are usually not met by the drainage material and a filter (geosynthetic or soil) is almost invariably required. It cannot be emphasized strongly enough that an adequate filter should be provided and that filter criteria should be met. Inadequate filtration is one of the most commonly encountered problems leading to failure of final covers, and yet the problem is easily solved by providing a properly designed and constructed filter.

2.3.2.2 Geosynthetic Materials. Geosynthetics, consisting of a drainage core and a geotextile filter, can be used to replace natural soil materials for the drainage layer. Of necessity, a geotextile filter must be used as a combined filter and separator, irrespective of the type of drainage core.

Regarding the *drainage core,* there are a wide variety of materials available, including:

- geonets of solid ribs with diamond-shaped apertures
- geonets of foamed ribs with diamond-shaped apertures
- "high flow" geonets of solid ribs in a parallel orientation
- drainage cores of single cuspations or dimples
- drainage cores of double cuspations or dimples
- drainage cores of built-up columns
- drainage cores of stiff three-dimensional nettings

Regarding the *geotextile filter,* there are also a wide variety of materials available, including:

- woven monofilament
- woven multifilament
- nonwoven needle-punched
- nonwoven heat-bonded

In general, the *geotextile filter* will be heat-bonded to the drainage core to eliminate a potentially low interface shear strength surface. Additionally, the heat-bonding keeps the geotextile in place so that fugitive soil particles cannot get into the apertures of the drainage core. Design details of both the drainage core and the geotextile filter will be provided later.

2.3.3 Design Details

Drainage design is well advanced for both granular soils and geosynthetics. Selected details follow.

2.3.3.1 Granular Materials. If composed of granular material such as sand or gravel, the drainage layer should meet the following specifications:

- Minimum thickness of 300 mm and minimum slope of 3% at the bottom of the layer; greater thickness and/or slope if necessary to provide sufficient drainage, as determined by site-specific design;
- Hydraulic conductivity of drainage material should be no less than 1×10^{-2} cm/sec (hydraulic transmissivity no less than 3×10^{-5} m^2/sec) at the time of installation;
- Should contain no debris or particles that could cause excessive damage to the underlying geomembrane, nor should it contain fines that might excessively clog or migrate within the layer to clog the outlet area; and
- A filter layer (granular or geosynthetic) should be included between the drainage layer and overlying protective soil layer to prevent excessive clogging of the drainage layer by fine particles.

The recommended 300 mm minimum thickness of the drainage layer allows sufficient thickness to avoid damage to underlying layers, such as a geomembrane. In some cases, particularly where unusually long drainage slopes may be part of the design, drainage layers thicker than 300 mm, hydraulic conductivities greater than 1×10^{-2} cm/s, and/or slopes greater than 3% may be necessary. The minimum value of 1×10^{-2} cm/sec for hydraulic conductivity was chosen because granular materials widely used as drainage media can provide this minimum hydraulic conductivity. In situations where the minimum criteria are insufficient or questionable, the design should utilize flow modeling in arriving at the flow-controlling design parameters.

The drainage layer must slope to an exit drain which allows percolated water to be efficiently removed. Examples of exit drains are shown in Figure 2.6. Care should be taken to provide adequate filtration around the drain, e.g., between gravel backfill around a perforated drainage pipe and sand employed in the drainage layer. The velocity of the exiting water should be controlled, within and beyond the exit drains, to prevent soil loss and destabilization. Large factors of safety may be needed to accommodate unexpected events. This may require an increase in drainage capability as one proceeds down the slope. This is illustrated in Figure 2.6. If the open channel in Figure 2.6c is used, the drainage stone at the toe must not become clogged with sediment. Also, consideration should be given to the toe drain during freezing weather. A frozen toe drain is a dysfunctional toe drain. If

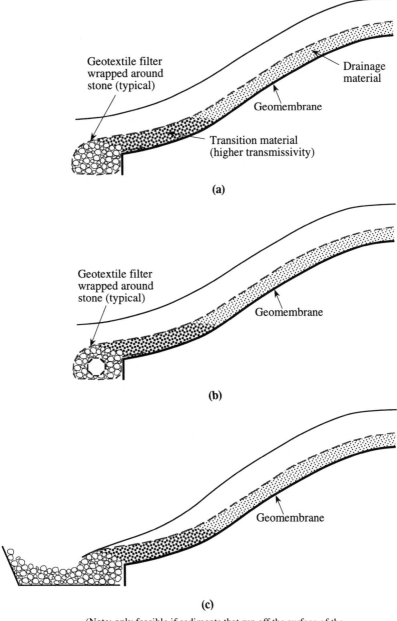

(Note: only feasible if sediments that run off the surface of the cover do not excessively clog the toe drainage material and that freezing cannot cause toe blockage.)

FIG. 2.6. *Various Designs Allowing for Free Drainage at Toe of Slopes [after Soong and Koerner (1996)]*

drainage is critical, it may be necessary to bury the toe drain to prevent it from freezing.

Materials used to construct the drainage layer may need to be washed or screened prior to construction to remove fines that may promote clogging. The design engineer should note that the stone may possess excessive fines even though the hydraulic conductivity of the stone is adequately large. This is because fines tend to migrate downslope, accumulate, and clog the drainage material near the critical outlet. If needed to prevent clogging of the drainage layer, a granular or geosynthetic filter should be placed directly over the drainage layer to minimize the migration of fines from the overlying topsoil into the drainage layer. If a graded granular filter is used, care should be taken to design the relationship of grain sizes according to the criteria presented below (Cedergren 1967).

$$\text{To prevent piping:} \qquad \frac{D_{15} \text{ (filter)}}{D_{85} \text{ (topsoil layer)}} < 4 \text{ to } 5, \text{ and} \qquad (2.4a)$$

$$\frac{D_{15} \text{ (drainage layer)}}{D_{85} \text{ (filter)}} < 4 \text{ to } 5 \qquad (2.4b)$$

$$\text{To maintain permeability:} \qquad \frac{D_{15} \text{ (filter)}}{D_{15} \text{ (topsoil layer)}} > 4 \text{ to } 5, \text{ and} \qquad (2.5a)$$

$$\frac{D_{15} \text{ (drainage layer)}}{D_{15} \text{ (filter)}} > 4 \text{ to } 5 \qquad (2.5b)$$

These criteria are cited by the Army Corps of Engineers for selection of a filter layer in relation to a soil to prevent the soil from piping through the filter. D_{85} refers to the particle size at which 85%, by dry weight, of the particles are smaller. D_{15} refers to the size below which only 15%, by dry weight, is finer. The criteria should be satisfied for all layers or media in the drainage system, including protected soil, filter material, and drainage material.

2.3.3.2 Geosynthetic Materials. If composed of geosynthetic materials, the drainage layer material should meet the following specifications:

- Same minimum flow capability as a granular drainage layer in the same situation; hydraulic transmissivity no less than 3×10^{-5} m²/sec under anticipated overburden for the design life;
- Inclusion of a geosynthetic filter layer above the drainage material to prevent intrusion and clogging by the overlying protective soil; and
- Inclusion of geosynthetic bedding beneath the drainage layer, if necessary, to increase friction and minimize slippage between

the drainage layer and the underlying geomembrane, and to prevent intrusion, by deformation, of the geomembrane into the geonet or drainage core of the drainage layer.

Since the normal stresses in a drainage system within the final cover are quite low (construction equipment is probably the largest), a wide range of geonets and geocomposites can be used for the drainage layer. Such geosynthetic drains would be an alternative to the granular soil drains just discussed. However, all geosynthetic drains require a geotextile filter acting as a separator to keep the protection soil from directly moving into the apertures of the geonet or drainage core. Furthermore, if the underlying barrier layer contains a geomembrane (as it usually does), a geotextile may have to be on the underside of the geonet or drainage core, as well as above it, to protect the geomembrane from puncture by the drainage geosynthetic.

The design of a geonet or geocomposite drainage core is straightforward. It results in the quantification of a flow rate factor of safety as follows:

$$FS = \frac{q_{allow}}{q_{reqd}} \qquad (2.6)$$

where

FS $\quad$ = factor of safety (to handle unknown hydraulic conditions or uncertainties),
q_{allow} = allowable flow rate, as obtained from laboratory testing, and
q_{reqd} = required flow rate, as obtained from design requirements of the actual system.

The allowable flow rate comes from laboratory testing of the product under consideration. The test setup must simulate the actual field system as closely as possible. If it does not model the field system accurately, then some adjustments to the laboratory value must be made. This is usually the case. Thus, the laboratory-generated flow rate is an ultimate value which must be reduced before use in design; that is,

$$q_{allow} < q_{ult} \qquad (2.7)$$

One way of doing this is to ascribe reduction factors to each of the items not simulated in the laboratory test. This can be accommodated as follows:

$$q_{allow} = q_{ult} \left[\frac{1}{RF_{IN} \times RF_{CR} \times RF_{CC} \times RF_{BC}} \right] \qquad (2.8)$$

or if all of the reduction factors are lumped together,

$$q_{allow} = q_{ult} \left[\frac{1}{\Pi RF} \right] \qquad (2.9)$$

where

q_{ult} = flow rate determined from a short-term transmissivity test between solid plates, e.g., ASTM D4716,

q_{allow} = allowable flow rate to be used for final design purposes,

RF_{IN} = reduction factor for elastic deformation, or intrusion, of the adjacent geosynthetics into the drainage core space,

RF_{CR} = reduction factor for creep deformation of the drainage core and/or adjacent geosynthetics into the drainage core space,

RF_{CC} = reduction factor for chemical clogging and/or precipitation of chemicals in the geonet's core space,

RF_{BC} = reduction factor for biological clogging in the drainage core space, and

ΠRF = product of all reduction factors for the site-specific conditions.

Other reduction factors, such as installation damage, temperature effects, and liquid turbidity, might also have to be included. If needed, they can be included on a site-specific basis. On the other hand, if the test procedure has included the particular item, the reduction factor would appear in the foregoing formulation as a value of unity. Details of the design and guidelines for various reduction factors are given in Koerner (1994).

2.3.3.3 Geotextile Filters.

2.3.3.3 Geotextile Filters. As noted previously, a geotextile must cover the geonet or drainage core, and its primary function will be to serve as a filter. In so doing, the geotextile must allow the water to pass without building up pore water pressure and, simultaneously, must retain the upstream soil so that upgradient piping and downgradient clogging of the geonet or drainage core do not occur. Thus, the design is a two-step process; first, for permeability (or permittivity); second, for soil retention (or apparent opening size).

Geotextile permeability is the first part of a geotextile filter design. It formulates a factor of safety using permittivity, which is the permeability divided by the geotextile's thickness, as follows:

$$FS = \frac{\psi_{allow}}{\psi_{reqd}} \qquad (2.10)$$

$$\psi = \frac{k_n}{t} \qquad (2.11)$$

where

ψ = permittivity,
k_n = cross-plane permeability coefficient, and
t = thickness at a specified normal pressure.

The testing for geotextile permittivity follows similar lines as used for testing soil permeability. In the United States, it follows ASTM D4491. Alternatively, some designers prefer to work directly with permeability and require the geotextile's permeability to be some multiple of the adjacent soil's permeability (e.g., 1.0 to 10.0, or higher).

The second part of a geotextile's filter design is focused on adequate upstream soil retention. There are many approaches to accomplishing a soil retention design, most of which use the upstream soil particle size characteristics and compare them to the 95% opening size of the geotextile (i.e., defined as O_{95} of the geotextile). The test method used in the United States to determine this value is called the apparent opening size (AOS) test. The AOS is defined as the approximate largest soil particle that would effectively pass through the geotextile. In Canada and Europe, the test method is called filtration opening size (FOS), and is accomplished by hydrodynamic sieving. It is felt by the authors to be the preferred method.

The simplest of the design methods examines the percentage of soil passing the No. 200 sieve, whose openings are 0.074 mm.

1. For soil with ≤50% passing the No. 200 sieve: $O_{95} < 0.59$ mm (i.e., AOS of the fabric ≥ No. 30 sieve)
2. For soil with >50% passing the No. 200 sieve: $O_{95} < 0.30$ mm (i.e., AOS of the fabric ≤ No. 50 sieve)

Alternatively, a series of direct comparisons of geotextile opening size (O_{95}, O_{50}, or O_{15}) can be made to some soil particle size to be retained (d_{90}, d_{85}, d_{50} or d_{15}). The numeric value depends on the geotextile type, the soil type, the flow regime, etc. For example, Carroll (1983) recommends the following widely used relationship

$$O_{95} < (2 \text{ or } 3)d_{85} \qquad (2.12)$$

where

d_{85} = soil particle size in mm for which 85% of the soil is finer, and
O_{95} = the 95% opening size of the geotextile.

Details of the design and example problems are given in Koerner (1994) and Koerner et al. (1995a).

2.4 HYDRAULIC/GAS BARRIER LAYER

No layer in the final cover cross-section is as important as the barrier layer, and no layer prompts as much controversy and differences of opinion. This section presents the authors' opinions on the subject, but it is fully recognized that site-specific, and most importantly waste-specific, conditions will dictate the type of materials used as a barrier layer and their configuration. What is clear is the different material categories from which to choose. They are:

- geomembranes (GMs),
- geosynthetic clay liners (GCLs), and
- compacted clay liners (CCLs).

Although other materials have been used (e.g., asphalt, as discussed previously), virtually all hydraulic/gas barrier layers are composed of one or more of the three materials listed above. These can be used by themselves as single barrier materials, or in composite form. Choices in the composite category are GM/GCL, GM/CCL or GM/GCL/CCL.

2.4.1 Geomembranes

Geomembranes form an essential part of most barrier layers. Of the factory-manufactured, polymeric geomembranes commercially available, the most common types being used in final covers are the following:

- High density polyethylene (HDPE)
- Very flexible polyethylene (VFPE) [this classification includes linear low density polyethylene (LLDPE), low density linear polyethylene (LDLPE), and very low density polyethylene (VLDPE)]
- Coextruded HDPE/VFPE/HDPE
- Flexible polypropylene (fPP)
- Polyvinyl chloride (PVC)

All of these geomembranes are available with smooth and textured surfaces for increased friction and shear strength when used on steep side slopes. They are also available as single-sided textured geomembranes, the other surface being smooth. Additionally, spray-on elastomeric geomembranes are possible, as are bituminous geomembranes. However, these groups are rarely used in final covers in comparison to the previously itemized geomembranes.

The general criteria one would expect that a geomembrane in a final cover should satisfy are the following:

- Provide a hydraulic barrier against water entering the waste or the lower permeability soil layer (CCL or GCL) beneath it.
- Provide a gas barrier preventing rising methane or other lighter-than-air gases from escaping to the atmosphere.
- Function as a hydraulic and/or gas barrier for the lifetime of the facility and its post-closure care period (and perhaps beyond).
- Accommodate the site-specific total settlement and differential settlement without failing.
- Provide for intimate contact with the underlying CCL or GCL, if being used as a composite liner.
- Provide adequate frictional resistance on its surfaces to adjacent materials so that slope stability is assured.
- Be capable of reasonable installation and seaming within the context of competing geomembranes and/or other barrier materials.
- Be of reasonable cost and availability with respect to other competing geomembranes and/or other barrier materials.

2.4.1.1 Permeation through Geomembranes. Regarding the hydraulic- and gas-barrier aspect of geomembranes, the permeation mechanism is one of vapor diffusion. This mechanism is very slow and tedious in comparison to hydraulic conductivity, which occurs when water flows through the voids of soil. Some vapor diffusion rates for geomembranes from the literature are as follows:

- 1.0 mm thick HDPE; water vapor rate $\simeq 0.020$ g/m^2–day
- 1.0 mm thick HDPE; solvent vapor rate = 0.20 to 20 g/m^2–day (depends on solvent type)
- 0.75 mm thick PVC; water vapor rate $\simeq 1.8$ g/m^2–day

Note that 1.0 g/m^2–day $\simeq 10$ l/ha–day, thus, all of these diffusion rates are very low. However, it is of far greater importance to install the geomembranes without punctures, tears, open seams, etc. Flow through such holes in a geomembrane will dwarf the diffusion values listed above.

2.4.1.2 Out-of-Plane Subsidence Behavior. Regarding total and differential settlement of the waste mass beneath the geomembrane, it is assumed that total settlement is generally a nonissue since deformation of a crowned cover system will not produce tensile stresses. Differential settlement is quite another matter. In this situation, tensile stresses will be generated. Obviously, the geomembrane must be able to accommodate the tensile stresses that are produced. Axisymmetric, out-of-plane, tests on var-

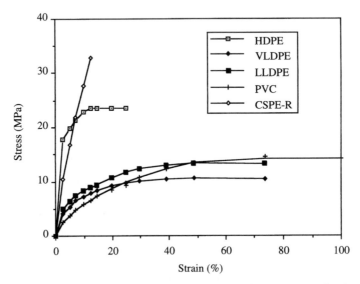

FIG. 2.7. Three-Dimensional Axisymmetric Stress-versus-Strain Response Curves of Various Types of Geomembranes

ious geomembranes have resulted in the stress-strain curves of Figure 2.7, after Koerner et al. (1990). Here it is seen that the accommodation of different geomembranes to differential settlement is poorest for scrim reinforced chlorosulphonated polyethylene (CSPE-R) and HDPE, and best for VLDPE, LLDPE and PVC. Note that VLDPE and LLDPE are both in the VFPE category. If large differential settlement is anticipated, as with municipal solid waste, the use of the high out-of-plane deformation geomembranes is recommended.

2.4.1.3 Interface Friction Using Textured Geomembranes.   Regarding frictional resistance of the geomembranes themselves, the advent of texturing has essentially eliminated low shear-strength surfaces. There are a number of manufacturing methods available to provide such texturing:

- co-extrusion for blown film manufacturing
- impingement for flat die manufacturing
- lamination for flat die manufacturing
- structuring via a heated calendar for flat die manufacturing

All of these processes result in increases of interface friction of 10 to 20 deg. against a well-graded sand. However, product-specific interface shear tests are always recommended.

2.4.1.4 Constructability. Regarding constructability of geomembranes, much has been written and discussed, e.g., see Daniel and Koerner (1996). More on this issue will be addressed in Chapter 7. It will suffice to say that geomembrane installation procedures (particularly field-seaming of rolls) have become highly developed and sophisticated, as have construction quality assurance procedures.

2.4.1.5 Cost and Availability. Regarding cost and availability of geomembranes, both are site-specific. The geomembrane industry is rapidly maturing with plants and facilities in nearly all industrialized countries of the world and a growing presence in emerging countries.

2.4.2 Geosynthetic Clay Liners

Geosynthetic clay liners consist of factory-manufactured rolls of bentonite placed between geotextiles or adhesively bonded to a geomembrane. The bentonite is the low hydraulic conductivity (or permeability) component; the geosynthetics act as containment materials. The geosynthetics also provide manufacturers with the opportunity to stitch-bond, needle-punch, or adhesively bond the bentonite into a self-contained unit suitable for handling, transportation, and placement as a barrier material. The structuring also results in internal shear strength for use of GCLs on sloping surfaces. The use of GCLs in final covers as the sole barrier material, or as a composite barrier with an overlying geomembrane, is rapidly growing in frequency and acceptance. Three U.S. EPA reports are available on GCLs [see Daniel and Boardman (1993), Daniel and Estornell (1991), and Daniel and Scranton (1996)].

Bentonite is the critical component of GCLs and the one that gives rise to the material's extremely low hydraulic conductivity (permeability). Bentonite is a naturally occurring, mined clay mineral that is extremely hydrophilic. When placed in the vicinity of water (or even water vapor), the bentonite attracts water molecules into a complex configuration that leaves little free-water space in the voids. This significantly decreases the hydraulic conductivity. The resulting hydraulic conductivity of most sodium bentonite GCLs is in the vicinity of 1×10^{-9} to 5×10^{-9} cm/sec.

The GCL products are manufactured as shown in the schematic of Figure 2.8. This processing gives rise to the various GCL styles currently available:

- adhesively bound bentonite between two geotextiles
- stitch-bonded types with bentonite between the two geotextiles
- needle-punched types with bentonite between the two geotextiles
- adhesively bound bentonite onto a geomembrane

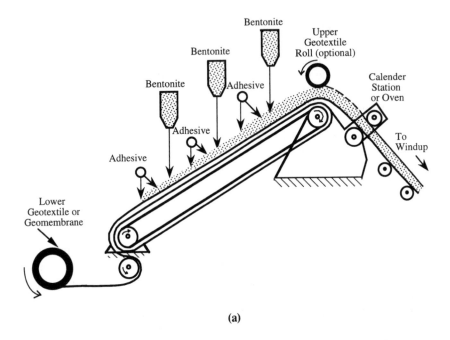

(a)

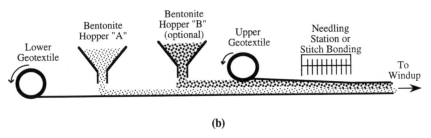

(b)

FIG. 2.8. *Schematic Diagram of Manufacture of Different Types of Geosynthetic Clay Liners*

The cross sections of these various styles are shown in Figure 2.9.

While the low hydraulic conductivity of GCLs gives rise to a favorable comparison of CCLs on the basis of a flow rate or flux calculation, the assessment of full technical equivalency is much more complicated. Koerner and Daniel (1994) have proposed the assessment of technical equivalency of GCLs to CCLs be made on the basis of numerous hydraulic, physical/mechanical, and construction issues. Table 2.1 presents the results of such a general assessment. Thus it is seen, in general, that GCLs are equivalent or superior to CCLs in final covers with the exception of certain

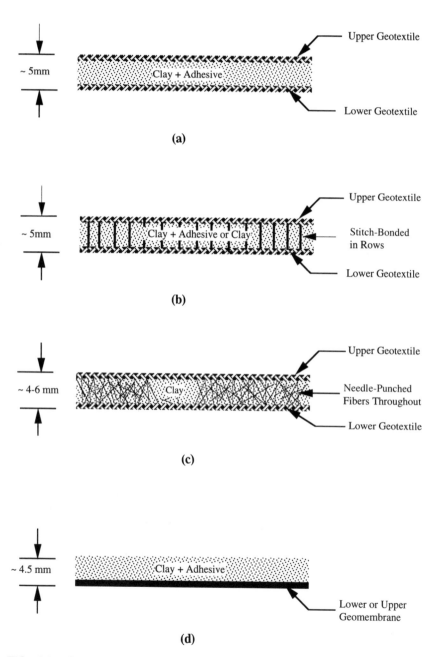

FIG. 2.9. Cross Section Sketches of Currently Available Geosynthetic Clay Liners

TABLE 2.1. Generalized Technical Equivalency Assessment for Final Covers

Category	Criterion for Evaluation	GCL is Probably Superior to CCL	GCL is Probably Equivalent to CCL	GCL is Probably Not Equivalent to CCL	Equivalency of GCL to CCL Depends on Site or Product
Hydraulic Issues	Steady flux of water		X		X
	Breakout time of water		X		
	Horiz. flow in seams or lifts	X			
	Horiz. flow beneath geomembranes	X			
	Generation of consolidation water				X
	Permeability to gases				
Physical/Mechanical Issues	Freeze-thaw behavior	X			
	Wet-dry behavior	X			
	Total settlement		X		
	Differential settlement	X			
	Slope stability			X	X
	Vulnerability to erosion			X	X
	Bearing capacity			X	
Construction Issues	Puncture resistance	X			
	Subgrade condition	X			
	Ease of placement	X			
	Speed of construction	X			
	Availability of materials	X			
	Requirements for water				
	Air pollution concerns				
	Weather constraints		X		
	Quality assurance considerations				X

field installation issues. However, with proper subgrade preparation and soil covering in a timely manner (i.e., before the GCL hydrates) and of sufficient thickness, *GCLs are the preferred clay barrier material for most final covers,* assuming slope stability is ensured.

The greatest design issue still outstanding regarding GCLs is the question of their shear strength when they are placed on side slopes of 3(H)-to-1(V), or steeper. There are three interface surfaces of concern: upper, internal, and lower. To investigate this situation at slopes of 3(H)-to-1(V) and 2(H)-to-1(V), full scale final cover test plots have been constructed in Cincinnati, Ohio. The ongoing study has been reported by Koerner et al. (1997) with the following preliminary findings:

- Those GCLs with woven slit film geotextiles on one or both surfaces must be carefully evaluated to be sure that hydrated bentonite does not extrude and lubricate the adjacent material interface (upper and/or lower) and cause unacceptable decreases in shear strength. Carefully modeled laboratory simulated direct shear tests are necessary to assess each site-specific and material-specific situation.
- Those GCLs relying on their dry (or as-manufactured) shear strength for stability must be fully and completely protected against hydration. This is usually accomplished by having geomembranes on both surfaces and, if so, the construction and deployment of the field-placed geomembrane must be flawless.
- Internal shear strengths of needle-punched and stitch-bonded GCLs appear to be stable at the 2(H)-to-1(V) slopes. If this trend continues, one can state that the factor of safety against internal shear failures of these particular GCLs is at least 1.0. This implies that the factor of safety of the same GCLs on the 3(H)-to-1(V) slopes is at least 1.5. This latter value is generally considered to be adequate.

Constructability issues involving GCLs will be described in Section 2.4.4.2. In that section, the discussion will center on composite liners, i.e., GM/GCL composites. Installation is the same, however, when a GCL is used by itself.

2.4.3 Compacted Clay Liners (CCLs)

Compacted clay liners are constructed primarily from natural soil materials that are rich in clay, although the liner may contain processed materials such as bentonite. Compacted clay liners are constructed in layers call *lifts* that typically have a thickness after compaction of 150 mm. On side slopes, the lifts in final cover systems are almost always placed parallel to the slope. However, parallel lifts are very difficult or impossible to construct on side slopes steeper than 2.5(H)-to-1(V).

2.4.3.1 Materials.

The recommended characteristics of materials used to construct CCLs in final cover systems that must have a hydraulic conductivity $\leq 1 \times 10^{-7}$ cm/s are as follows:

Percentage of Fines:	$\geq$30 to 50%
Plasticity Index:	$\geq$7% to 15%
Percentage of Gravel:	$\leq$20% to 50%
Maximum Particle Size:	25 to 50 mm

Percentage of fines is described as the percent by dry weight passing the No. 200 sieve, which has openings of 0.074 mm. Plasticity index, which is defined as the liquid limit minus plastic limit, may be determined by ASTM D4318. Percentage of gravel is defined as the percent, by dry weight, retained on a No. 4 sieve (4.76 mm openings). Local experience may dictate more stringent requirements, and, for some soils, more restrictive criteria may be appropriate. However, if the criteria tabulated above are not met, it is unlikely that a natural soil liner material will be suitable without additives such as bentonite.

Compacted clay liners used in final cover systems must be as ductile as possible (to accommodate differential settlement) and must be resistant to cracking from moisture variations e.g., desiccation. Sand-clay mixtures are ideal materials if resistance to shrinkage and desiccation-induced cracking are important (Daniel and Wu 1993). Ductility is achieved by avoiding the use of dense, dry soils, which tend to be brittle.

If suitable materials are unavailable, local soils can be blended with commercial clays, e.g., bentonite, to achieve low hydraulic conductivity. A relatively small amount of sodium bentonite (typically 2 to 6% by weight) can lower hydraulic conductivity as much as several orders of magnitude. The percent bentonite is usually defined as the weight of bentonite (including a small amount of hydroscopic water) divided by the weight of soil (dry and moist weight have been used, but the dry weight is recommended) to which bentonite is added times 100%. Soils with a broad range of grain sizes usually require a relatively small amount of bentonite ($\leq$6%). Uniform-sized soils, such as dune sand, usually require more bentonite (up to 10 to 15%). Sometimes materials are blended to provide a material with a broad range of grain sizes that minimizes bentonite content. For instance, on one project, a coarse-to-medium sand was successfully blended with bentonite. By adding 30% of fine, inert material (waste fines from a materials processing plant), the amount of bentonite required was halved.

2.4.3.2 Compaction Requirements.

The objective of compaction is to densify the soil (hydraulic conductivity decreases with decreasing porosity), and to remold chunks (*clods*) of soils into a homogeneous mass that is free of large, continuous interclod voids. If these objectives are accomplished with suitable soil materials, low hydraulic conductivity, e.g., less than 1×10^{-7} cm/s, will result. Whether the as-built low hydraulic conductivity is maintained is a separate question. In fact, it is easier to build a low-hydraulic-conductivity CCL than it is to design a final design cover system that will adequately protect the CCL from forces (e.g., differential settlement and desiccation) that tend to drive the conductivity above the design value.

Experience has shown that the water content of the soil, method of compaction, and compactive effort have a major influence on the hydraulic conductivity of compacted soil liners. Laboratory studies have demonstrated that low hydraulic conductivity is easiest to achieve when the soil is compacted wet of optimum water content with a high level of kneading-type compactive energy. The soil must be sufficiently wet so that, upon compaction, clods of clayey soil will mold together, eliminating large interclod pores. However, the soil must not be too wet. Excessively wet soils are extremely vulnerable to damage from desiccation. The approach described by Daniel and Wu (1993) is recommended for establishing appropriate water content-density criteria that will ensure both low as-built hydraulic conductivity and good resistance to desiccation cracking.

2.4.3.3 Construction.
The field construction of compacted clay liners is obviously critical to their proper functioning.

2.4.3.3.1 Processing. Some liner materials need to be processed to break down clods of soil, to sieve out stones and rocks, to properly wet the soil, or to incorporate additives. Clods of soil can be broken down with tilling equipment. Stones can be sieved out of the soil with large vibratory sieves, mechanized "rock pickers" passed over a loose lift of soil, or by laborers who remove oversized material by hand. Pulverization machines can process soil in a loose lift, breaking down hard chunks and crushing stones or large clods.

Additives such as bentonite can be introduced in two ways. One technique is to mix soil and the additive in a pugmill, which is the type of equipment used to mix the ingredients of concrete. Alternatively, the soil can be spread in a loose lift 200–300 mm thick, the additive spread over the surface, and rototillers used to mix the materials. Several passes of the tiller over a given spot are usually needed. Of the two methods, the pugmill is more reliable in providing thorough, controlled mixing.

2.4.3.3.2 Surface Preparation. It is crucial that each lift of a soil liner be effectively bonded to the overlying and underlying lifts. The surface of a previously-compacted lift should be rough rather than smooth. Many contractors like to smooth roll the completed lift of soil with a smooth steel-drummed roller. The smooth-rolled surface is desirable to promote runoff from rainstorms (a rough surface holds water in tiny puddles and it may take several days to dry out the soil so that construction can resume) and to provide a hard skin that minimizes desiccation. However, if the next lift is placed on a hard, smooth surface, a distinct lift interface will develop, and the interface provides horizontal hydraulic connection between permeable zones in adjacent lifts. On the other hand, if the surface is rough, the new and old lifts blend into one another. Discs are used to roughen (*scarify*) the surface of a previously-compacted lift to a depth of about 25 mm.

2.4.3.3.3 Placement. Soil is placed in a loose lift that is no thicker than about 230 mm. If grade stakes are used to gauge thickness, the stakes should be removed and the hole left by the stakes sealed. Other techniques that avoid use of stakes, e.g., use of lasers, are preferable for control of elevations. After the soil is placed, a small amount of water may be added to offset evaporative losses, and the soil may be tilled one last time prior to compaction.

2.4.3.3.4 Compaction. Heavy, compactors with large feet that fully penetrate a loose lift of soil are ideal. Rollers with feet that fully penetrate a loose lift of soil pack the base of a new lift into the surface of the previously-compacted lift, which helps to bond lifts together. The long feet also help to break down and remold clods of soil over the full thickness of a lift. Recommended compactor specifications include a minimum mass of 18,000 kg and minimum foot length of 180 to 230 mm (but the foot should have a length no smaller than the thickness of a loose life). However, in many landfill covers it is simply not possible to use such heavy compactors because the foundation (underlain by waste at shallow depth) may not be adequate to support the weight of the equipment. Lighter-than-ideal equipment must be used in such cases. To compensate for the light weight, it may be necessary to use thinner lifts and more passes of the compactor.

Static (dead weight) compactors are preferred over vibratory compactors for soil liners. The weight of the compactor must be compatible with the soil: relatively dry soils with firm clods require a heavy compactor whereas relatively wet soils with soft clods require a roller that is not so heavy that it becomes bogged down in the soil. Also, it is sometimes desirable to compact the lift with two compactors. A heavy roller with fully penetrating feet compacts the soil initially. If this roller leaves loose material in the upper part of the lift, a roller with short feet (*pad foot* roller), rubber-tired equipment, or a smooth steel-drum roller can be used to compact the upper part of a lift. It may be particularly important to have more than one type of compactor available on landfill cover projects because the strength of the supporting foundation may be variable and unpredictable.

Soil-bentonite liners can often be compacted with rubber-tired or smooth-drum rollers. Soil-bentonite mixtures do not develop clods, and densification of the soil is often the primary objective with soil-bentonite liners. However, rollers with fully penetrating feet may be effective in bonding soil-bentonite lifts.

2.4.3.3.5 Protection. After compaction of a lift, the soil should be protected from desiccation and freezing. Desiccation can be minimized in several ways: the lift can be temporarily covered with a sheet of plastic (but one must be careful that the plastic does not heat excessively and thereby dry the clay), the surface can be smooth-rolled to form a relatively impermeable layer at the surface, or the soil can be periodically moistened. The completed CCL can be covered with a thin (100 to 200 mm thick) layer of soil that is kept moist to prevent desiccation. The protective soil is then stripped away just before the next layer, e.g., a geomembrane, is installed. The compacted lift can be protected from frost damage by avoiding construction in freezing weather or by temporarily covering the lift with an insulating layer.

2.4.3.3.6 Quality Assurance. Quality control and assurance are critical. Procedures are described by Daniel and Koerner (1996).

2.4.3.4 Test Pads. The construction of a test pad prior to building a full-sized liner has many advantages. By constructing a test pad, one can experiment with such variables as compaction water content, construction equipment, number of passes of the equipment, lift thickness, etc. Most importantly, though, one can conduct extensive testing, including quality control testing and in-situ hydraulic conductivity testing, on the test pad to verify performance criteria and the effectiveness of proposed QA procedures.

It is usually recommended that the test pad have a width of at least 3 construction vehicles and an equal or greater length. The pad should ideally be the same thickness as the full-sized liner, but the test pad may be thinner. (The full-thickness liner should perform at least as well as, and probably better than, a thinner test pad.) The in-situ hydraulic conductivity may be determined in many ways. The sealed double-ring infiltrometer is usually the best large-scale test (Daniel 1989; and ASTM D5093), although the Boutwell test (Trautwein and Boutwell 1994) is enjoying increased popularity due to its ease of operation and relatively short testing times.

2.4.3.5 Shear Strength. The shear strength of the CCL, and particularly the CCL-geomembrane interface, can be critical to the stability of the final cover system. Slope stability is discussed in Chapter 5. The designer should realize that low hydraulic conductivity is most easily achieved by adding water to the clay, thereby compacting the clay wet of its optimum water content. However, the conditions that tend to result in a low hydraulic conductivity in the CCL also tend to cause low internal and interfacial

shear strength. The selection of appropriate water content/density parameters is usually a compromise between the need for low hydraulic conductivity and the need for adequate shear strength. The designer simply cannot
become obsessed with hydraulic conductivity to the extent that inadequate
attention is given to the shear strength of the CCL and CCL interfaces with
other materials.

2.4.4 Composite Liners

The authors believe that a composite liner generally provides the preferred barrier layer for an environmentally safe and secure final cover. Recognize that, in most cases, the barrier layer must function both as a liquid
barrier to potentially downward-moving water and simultaneously as a gas
barrier to upward-moving gases. It is felt that the preferred combination for
this composite liner is a geomembrane over a geosynthetic clay liner (GM/
GCL). For MSW and abandoned dumps, the geomembrane component will
generally be VFPE, fPP, or PVC due to the excellent out-of-plane deformation characteristics of these materials. For hazardous and radioactive
wastes which are relatively stable, the geomembrane component will generally be HDPE due to its excellent durability and subsequent long lifetime.
The GCL component will generally be internally reinforced by needle-
punching or stitch-bonding, although if slope stability is not a problem,
other types of GCLs are commonly used. The characteristics of the geomembranes and GCLs tend to compliment one another, so that the long-
term effectiveness of the two components together is greater than each
alone. In short, the geomembrane will tend to roof over the inconsistencies
in the underlying GCL, while the GCL will tend to plug any leakage through
a hole in the overlying geomembrane. In addition, each component tends
to back up the other in the event of a failure of either.

*It should be noted that this recommendation of the use of a GM/GCL
composite is a major departure from many federal and state regulations
which tend to recommend a GM/CCL as the preferred composite barrier.*

The authors believe that a GM/CCL composite is *not the best selection*
for most final cover systems, and, had GCLs been available at the time of
writing of many of these regulations, the choice would have been a GM/
GCL. Additional arguments for this preference will be provided later.

2.4.4.1 *Geomembrane Component.*   In no case should the thickness
of the geomembrane be less than 1.0 mm. Note that this value is intermediate between regulations in the United States and in Germany. The
authors believe that this is the minimum acceptable thickness to meet cover
objectives and still be sufficiently rugged to withstand field seaming and
the expected stresses during construction backfilling and operations. The
adequacy of the selected thickness should be demonstrated by an evalua-

tion considering the type, strength, and durability of the proposed geomembrane material, its type of seaming, and site-specific factors such as: steepness of slopes, physical compatibility with the material used for the underlying and overlying layers, stresses during installation, expected overburden, climatic conditions, settlement, and subsidence.

Geomembrane failure mechanisms have been discussed in the open literature. Most failures appear to result from inadequacies in either some design assumption or in installation and backfilling. It follows then that most failures can be prevented if (a) a peer review of the design is undertaken and (b) a strict quality assurance program is adhered to during the construction process. Great emphasis on construction quality assurance, particularly in the construction of barrier layers, should be undertaken. The CQA personnel themselves should be certified, and at least one such program is available (see Daniel and Koerner 1996).

One of the causes of geomembrane failure in the lining systems **beneath** landfills and surface impoundments is chemical incompatibility. However, the geomembrane in a final cover system should not come into direct contact with any wastes and chemical incompatibility should not be of concern, particularly if a gas collection layer is present. This makes it possible to accept a wider range of geomembrane materials in cover systems above the waste than in liner systems below the waste. There is no overriding technical reason that the bottom liner and cover barrier necessarily be constructed of the same material.

As already mentioned, one of the primary causes of geomembrane failure is damage during installation, backfilling, or operations. To aid in preventing damage, such as punctures, rips, and tears, at least 300 mm and preferably 450 mm of bedding material above and below the geomembrane is recommended. Since the geomembrane is in direct contact with an underlying GCL or CCL, that layer will serve as the geomembrane bedding. In most cases, the drainage layer above the geomembrane will suffice as the overlying geomembrane bedding, if it is of natural soil. If it is a geosynthetic, a minimum thickness of 300 mm of soil above the drainage geocomposite is necessary. The actual bedding thickness should, however, be based upon consideration of failure mechanisms and construction methods potentially harmful to the geomembrane, e.g., if construction equipment or methods are capable of penetrating the drainage layer and tearing, ripping, or puncturing the geomembrane, then the thickness should be increased. Truck traffic can be more damaging than the stresses imposed by construction placement equipment. Field test pads should be constructed if this issue is a concern. If the design thicknesses for drainage and bedding layers differ, then the greater thickness should be used.

Penetration of the geomembrane by gas vents, condensate drainage pipes, or leachate recirculation pipes should be minimized. Where a penetration is necessary, it is essential to obtain a secure, liquid-tight seal between the structure and the geomembrane to prevent leakage of cover soil water around the vent. Settlement and/or lateral movements of the materials around the vent may create tearing stresses in the geomembrane, which should be taken into account in the design of both the vents and the associated geomembrane pipe boots.

Differential settlement of the underlying waste will cause out-of-plane induced tensile stresses that must be accounted for in the geomembrane design. Care should be taken to accommodate for these and other stresses. For example, excess slack might be created intentionally to reduce stress, but then may, in turn, result in stresses in the folds that can lead to long-term failure of the geomembrane. Differential settlement and/or lateral movement are the primary reasons why flexibility should be considered in the design and selection of the geomembrane type (recall Figure 1.7a).

The subgrade for the geomembrane must be carefully prepared and smoothed so that no small-scale stress points are created due to protrusions of rocks or other materials. In most cases, this should cause no difficulty, since the subgrade will be a GCL or CCL.

Field-seaming of the geomembrane must be done carefully by technicians qualified and experienced in seaming the particular geomembrane being installed. Thermal seaming methods, e.g., hot wedge, hot air, extrusion and ultrasonics, are recommended over all other types of geomembrane seaming. Particular care is required during the surface preparation, i.e., grinding, phase of extrusion seaming. Holes can result from poorly made seams or those not sufficiently sturdy to withstand unavoidable stresses. All geomembranes will expand and shrink with temperature changes. These characteristics may promote later leakage if not carefully considered in the construction process. All of the potential failure causes can be minimized or prevented by using expert installers and adhering to a strict construction quality assurance program.

2.4.4.2 Geosynthetic Clay Liner Component.
The geosynthetic clay liner (GCL) component of the composite liner is placed on the underlying gas collection layer or foundation layer. The subgrade should be carefully prepared with no stones on the surface greater than 12 mm. Frozen ruts are to be avoided. The GCLs arrive in rolls placed directly on the prepared subgrade. If there is a preferred orientation (top or bottom), it must be properly constructed. The edges and ends of GCL rolls are handled by making an overlap of 150 to 200 mm (more if differential settlement might cause a reduction in the width of overlaps) and shingling in a downgradient direction. When the geotextiles on top and/or bottom of the GCL are needle-punched nonwovens, a layer of dry or moist bentonite should be placed in the overlapped area.

The GCLs should be protected against hydration until the covering geomembrane and at least 300 mm of soil is placed above. Note that if a drainage geocomposite is used, this material must be placed before the soil covering layer. Other than precipitation hydrating the GCL, water present in the subgrade soil can hydrate the bentonite within the GCL. Daniel et al. (1993) have shown that such hydration can occur in 10 to 15 days, even on relatively dry soil subgrades. The danger in driving over hydrated GCLs is their resulting loss of thickness due to lateral squeezing of the bentonite (Koerner and Narejo 1995). In extreme cases, wheel loads on an uncovered, hydrated GCL can completely compromise its thickness, squeezing bentonite laterally from beneath the loaded area.

While GCLs are straightforward and rapid to place, they still must be handled and positioned so that loss of bentonite does not occur. Fugitive bentonite has resulted in excessively clogged drainage layers, in difficulties while seaming geomembranes (particularly textured geomembranes), and in loss of thickness of the GCL itself. Good construction practices and quality assurance monitoring is essential in the placement of GCLs, as it is with other geosynthetics and natural soils.

2.4.4.3 Compacted Clay Liner Component.
A compacted soil component beneath, and in direct contact with, the geomembrane will:

- minimize, over the long term, liquid migration into the waste in the event of geomembrane failure or through imperfections (holes, tears, etc.) inadvertently left during the construction process,
- provide a firm foundation for the overlying layers of the cover system,
- serve as bedding material for protection of the overlying geomembrane, and
- in conjunction with the geomembrane, satisfy the typical regulatory requirement for the cover to be no more permeable than the bottom liner of the facility.

The design of the CCL will depend on site-specific factors including the properties and engineering characteristics of the soil being compacted, the degree of compaction attainable, the total expected load, and the expected precipitation. These issues were described previously in Section 2.4.3.

The unique aspect of construction of a CCL by itself versus a GM/CCL composite is that, in the latter case, the work is generally performed by two separate contracting organizations. The CCL is constructed by an earthwork contractor and the geomembrane by a geosynthetics installer. They rarely are the same organizations. Thus, timing and care represent ongoing concerns.

Regarding *timing*, it is ideal if the geomembrane installer is mobilized and ready to work immediately following completion of the clay liner by the earthwork contractor. As the final lift of compacted clay is placed and accepted, the geomembrane can be placed immediately. Unfortunately, this is rarely the case. All to often, days, weeks, or even months pass after completion of the clay liner and before geomembrane placement. During this gap in construction activity, the compacted clay liner must be protected against:

- drying and desiccation cracking,
- wetting and rutting by vehicles, and
- freezing.

These are difficult (and expensive) challenges, thus it behooves all parties to have the geomembrane placed as soon as possible.

Regarding *care*, the earthwork contractor must leave the site accessible to the geomembrane installer as well as conforming to proper line and grade. Having earthwork equipment putting final touches to the surface of the compacted clay while cornered or even surrounded by exposed and deployed geomembranes is an invitation to disaster. In a similar manner, the geomembrane installer must respect the work of the earthwork contractor in having provided proper line and grade. The handling equipment for transporting large rolls of geomembranes can leave deep ruts and irregularities in the surface of the compacted clay. Situations such as working too soon after a rainfall simply must be avoided. In cases of geomembrane placement in cold weather, the incidence of large frozen ruts in the surface of the clay liner is unacceptable. As with stones on the subgrade, the maximum depth (or height) of frozen ruts should be not more than 12 mm.

2.4.4.4 Intimate Contact Concern. Regarding intimate contact of a geomembrane with an underlying CCL or GCL, the incidence of waves or wrinkles is of concern. These waves form in the geomembrane after initial placement and subsequent heating during the day, along with direct exposure to sunlight. While more pronounced in the stiffer and thicker geomembranes like HDPE, the waves occur in all types of geomembranes since their expansion/contraction characteristics are largely the same (Koerner 1994).

At issue is not that waves occur in the geomembrane, but when backfilling occurs. At night, as the temperature is low and the sun is not shining, the geomembrane contracts and the waves are eliminated (provided too much slack is not installed in the seamed system). Thus backfilling should occur from daybreak until the waves begin to reform. At that time (typically 10:00 AM) backfilling should cease until the following morning. Of course, night backfilling is acceptable but has its own difficulties, e.g., accidents, worker safety, adequate lighting, higher costs, etc. It should also be noted that the use of white surfaced geomembranes (via co-extrusion) decreases wave heights by approximately one-half (Koerner et al. 1995b). Thus, sunlight exposure is less of a factor and backfilling can continue longer into the day than stated above.

For GCLs as the lower component of a composite liner, lateral transmission of liquid in the upper geotextile has been evaluated by Harpur et al. (1994) and found to be of little concern. Apparently, the hydrated bentonite fills in, or extrudes through, the voids of the geotextile, greatly decreasing the transmissivity within the contact area to the geomembrane. This, however, gives concern in another respect: the possibility of decreasing the shear strength of the geomembrane-to-GCL interface. Proper direct shear testing and slope stability analyses are required when this type of composite barrier is on steep slope angles.

2.5 GAS COLLECTION LAYER

The authors offer the following design recommendations for a gas collection layer located beneath the barrier layer and above the foundation layer or waste itself:

- When using natural soils, the layer should be a minimum of 300 mm thick.
- Geosynthetics can be used, if shown to have equivalent transmissivity characteristics.
- Materials used in construction of the gas collection layer should be coarse-grained, highly permeable soils (such as sands) or geosynthetics. The geosynthetics could be thick needle-punched nonwoven geotextiles, geonet composites, or related drainage geocomposites.
- When using natural soil gas collection layers, venting to an exterior collection point for disposal or treatment should be provided by means such as horizontal perforated pipes, patterned laterally throughout the gas vent layer, which channel gases to vertical risers.
- When using geosynthetics, piping systems are not needed since flow rates are considerably higher than with sandy soils.
- The number of vertical risers through the cover system should be minimized and located at high points in the cross section. They must be designed to prevent water infiltration through and around them.

An alternative design, particularly useful for layered landfills where vertical migration of gases is impeded, may include perforated vertical collector pipes penetrating deep into the landfill. In this case, several cover penetrations may be required, one for each standpipe. Here again, the penetrations should be securely sealed to the low-permeability layer. The standpipes may be 300 mm or more in diameter and may be dual purpose, e.g., to serve as leachate injection pipes for rapid waste degradation.

2.5.1 Design

The design of a gas collection layer, either natural soil or geosynthetic, should proceed so as to formulate a flow rate factor of safety (FS) as follows:

$$FS = \frac{q_{allow}}{q_{reqd}} \qquad (2.13)$$

where

q_{allow} = allowable (test) flow rate, and
q_{reqd} = required (design) flow rate.

For the value of "q_{allow}," a permeability test, now using gas as the permanent, can be configured for sand as well as geosynthetics. Data are available for needle-punched nonwoven geotextiles (Koerner 1994), and the same technique can be used for other materials. For the value of "q_{reqd}," an estimate of the gas release rate is required. For actively decomposing municipal solid waste materials, the rates can be very high. "Blow-outs" of geomembranes in final covers, completely displacing the overlying cover soils, have occurred (recall Figure 1.7b). Thus, very conservative factors of safety values are recommended. For non-degrading wastes, the flow rates can be very low. Thus, lower factor of safety values are appropriate.

Whatever the factor of safety, it must be emphasized that with a geomembrane in the barrier layer, rising gases of any type and flow rate simply cannot escape as they can with only a soil barrier layer by itself (either GCL or CCL). The gas collection layer is essential in both its existence and in its proper grading to outlet vents, which necessarily must penetrate the barrier, drainage, protection, and surface layers. Some ideas on this critical penetration detail for both CCLs and GCLs as well as their overlying geomembrane are shown in Figure 2.10.

2.5.2 Construction

Materials used in construction of the gas collection layer should have specifications similar to the granular material used for the drainage layer or to similar geosynthetic drainage materials. The materials should be placed in a way that facilitates the placement and compaction of the overlying

GCL or CCL. Once placed, the gas drainage layer should allow free movement of gases to collection pipes and/or outlet points,.

The outlets may consist of pipes or vents allowing the gas to be collected, vented, or treated. The vent layer and outlet should be designed to minimize cover penetrations which could allow possible liquid infiltration through the cover (recall Figure 2.10). Outlet vents should be constructed through the barrier layer at the highest elevation of the gas vent layer to allow maximum evacuation of gas.

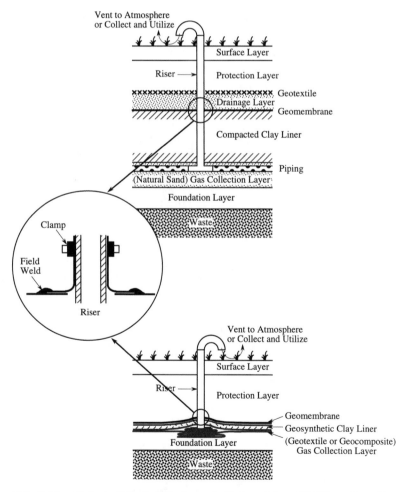

FIG. 2.10.   Selected Venting Systems to Transmit Landfill Gases from Gas Collection Layer (Natural Soil and Geosynthetics) to Ground Surface

In addition to providing gas removal, the gas collection layer will provide a protective foundation upon which to place a GCL or construct a CCL. The collection layer must be placed over the underlying waste to design elevation, allowing for estimated settlement, prior to placement and compaction of overlying layers. A filter, either granular or geotextile, may be required between the gas vent layer and the CCL to prevent excessive clogging. Typically, the lower geotextile of a GCL will be adequate to prevent downward extrusion of hydrated bentonite, but the situation must be addressed and the GCL appropriately specified.

Alternatively, vertical standpipe gas collectors are constructed of perforated sections, being built up as the unit is filled with waste. They may be constructed of concrete and wrapped with geosynthetic filter material to prevent clogging of the perforations. This geotextile must be carefully selected so as not to biologically clog over time (Koerner et al. 1995a).

2.6 FOUNDATION LAYER

The layer upon which the entire final cover cross section will be constructed is the foundation layer. Depending on the type of waste being covered, it can be the last lift of daily soil cover, a temporary soil cover, or a previously placed soil cover which does not meet regulatory standards.

Irrespective of the site-specific situation, the foundation layer represents the last time that mechanical compaction can take place so as to minimize final cover settlements from occurring. For this reason, the foundation layer is always heavily proof rolled with large compactors. As many load repetitions as practical are used so that stresses are felt as deeply as possible in the waste mass.

In extreme circumstances, deep dynamic compaction (DDC) has been used. In this method a large weight (usually a concrete block) is dropped from a great height (many meters), mobilizing a tremendous energy on the surface of the waste mass. The resulting craters are eventually filled and the surface is proof-rolled with compaction equipment. Galenti et al. (1991) describes the technique, which has been used at a number of landfills, particularly when post-closure use is to be made of the surface. The depth of influence of the technique can be estimated from the following equation:

$$D = 0.5(WH)^{1/2} \tag{2.14}$$

where

D = depth (in meters),
W = mass of the following weight (in metric tons), and
H = height of the following weight (in meters).

It has been estimated that for soil densification (not necessarily solid waste), the improvement is substantial down to $1/2D$, beyond which it decreases (Mayne et al. 1984).

As often is the case, when the upper 300 mm of the foundation layer is a granular soil, it can also act as the gas collection layer.

Lastly, it should be mentioned that elevations for all subsequently placed layers of the final cover follow the lines and grades of the finished surface of the foundation layer. These final lines and grades should be survey-controlled and verified by the on-site monitoring organization.

2.7 REFERENCES

Boschuk, J. J. (1991). "Landfill Covers: An Engineering Perspective," Geotechnical Fabrics Report, Vol. 9, No. 2, IFAI, pp. 23–34.

Carroll, R. G., Jr. (1983). "Geotextile Filter Criteria," TRR96, Engineering Fabrics in Transportation Construction, Washington, DC, pp. 46–53.

Cedergren, H. R. (1967). Seepage, Drainage, and Flow Nets, John Wiley & Sons, Inc., New York, NY.

Cline, J. F. (1979). "Biobarriers Used in Shallow-Burial Ground Stabilization," Pacific Northwest Laboratory Report for U.S. Department of Energy. PNL Report No. 2918.

Cline, J. F., Cataldo, D. A., Burton, F. G., and Skiens, W. E. (1982). "Biobarriers Used in Shallow Burial Ground Stabilization," Nuclear Technology, Vol. 58, August, pp. 150–153.

Daniel, D. E. (1989). "In Situ Hydraulic Conductivity Tests for Compacted Clay," Journal of Geotechnical Engineering, Vol. 115, No. 9, pp. 1205–1226.

Daniel, D. E., and Boardman, B. T. (1993). "Report of Workshop on Geosynthetic Clay Liners," EPA/600/R-93/171, U.S. Environmental Protection Agency, Cincinnati, OH, 106 pgs.

Daniel, D. E., and Estornell, P. M. (1991). "Compilation on Alternative Barriers for Liner and Cover Systems," EPA/600/2-1/002, Environmental Protection Agency, Cincinnati, Ohio.

Daniel, D. E., and Koerner, R. M. (1996). Waste Containment Facilities: Guidance for Construction Quality Assurance and Quality Control of Liner and Cover Systems, ASCE Press, New York, NY, 354 pgs.

Daniel, D. E., and Scranton, H. B. (1996). "Report of 1995 Workshop on Geosynthetic Clay Liners," U.S. Environmental Protection Agency, Cincinnati, Ohio, 96 pgs (in press).

Daniel, D. E., Shan, H.-Y., and Anderson, J. D. (1993). "Effects of Partial Wetting on the Performance of the Bentonite Component of a Geosynthetic Clay Liner," Proc. Geosynthetics '93, IFAI Publ., St. Paul, MN, pp. 1483–1496.

Daniel, D. E., and Wu, Y. K. (1993). "Compacted Clay Liners and Covers for Arid Sites," Journal of Geotechnical Engineering, Vol. 119, No. 2, pp. 223–227.

DePoorter, G. L. (1982). "Shallow Land Burial Technology Development," Report to the Low-Level Waste Management Program Review Committee. Los Alamos National Laboratory, Los Alamos, NM.

Galente, V. N., Eith, A. W., Leonard, M. S. W., and Finn, P. S. (1991). "An Assessment of Deep Dynamic Compaction as a Means to Increase Refuse Density for an Operating Landfill," Proc. Midland Geotechnical Society, United Kingdom, July, pp. 183–193.

Garrels, (1951). In J. K. Mitchell's, Fundamental of Soil Behavior, J. Wiley and Sons, New York, NY, 1976.

Gee, G. W., Fayer, M. J., Rockhold, M. L., and Campbell, M. D. (1992). "Variations in Recharge at the Hanford Site," Northwest Science, Vol. 66, pp. 237–250.

Harpur, W. A., Wilson-Fahmy, R. F., and Koerner, R. M. (1994). "Evaluation of the Contact Between GCLs and Geomembranes in Terms of Transmissivity," Proc. Conf. on Geosynthetic Liner Systems, R. M. Koerner and R. F. Wilson-Fahmy, Eds., IFAI, St. Paul, MN, pp. 143–154.

Hokanson, T. E. (1986). "Evaluation of Geologic Materials to Limit Biological Intrusion of Low-Level Radioactive Waste Disposal Sites," Los Alamos National Laboratory, Report No. LA-10286-MS.

Hsuan, Y. G., and Koerner, R. M. (1995). "Long-Term Durability of HDPE Geomembranes, Part I—Depletion of Antioxidants," GRI Report #16, December 11, 1995, Philadelphia, PA, 37 pgs.

Hudson, R. Y. (1959). "Laboratory Investigation of Rubble Mound Breakwaters," Proc. Waterways and Harbors Div., ASCE, September, pp. 93–121.

Johnson, D. I., and Urie, D. H. (1985). "Surface Barrier Caps: Long-Term Investments in Need of Attention," Waste Management and Research, 3, pp. 143–148.

Kemper, W. D., Nicks, A. D., and Corey, A. T. (1994). "Accumulation of Water in Soils under Gravel and Sand Mulches," Soil Science Society of America Journal, Vol. 58, pp. 56–63.

Koerner, G. R., Koerner, R. M., and Martin, J. P. (1995a). "Design of Landfill Leachate Collection Filters," Jour. of Geotechnical Engineering Div., ASCE, Vol. 120, No. 10, October, pp. 1792–1803.

Koerner, G. R., and Koerner, R. M. (1995b). "Temperature Behavior of Field Deployed HDPE Geomembranes," Proc. Geosynthetics '95, IFAI, pp. 921–937.

Koerner, R. M. (1994). Designing with Geosynthetics, Third Edition, Prentice-Hall, Englewood Cliffs, NJ, 783 pgs.

Koerner, R. M., Carson, D. A., Daniel, D. E., and Bonaparte, R. (1997). "Current Status of the Cincinnati GCL Test Plots," Proc. GRI-10 Conference on Field Performance of Geosynthetics, GII Publ., Philadelphia, PA, pp. 153–182.

Koerner, R. M., and Daniel, D. E. (1994). "A Suggested Methodology for Assessing the Technical Equivalence of GCLs to CCLs," Proc. GRI-7 Conference on Geosynthetic Liner Systems, IFAI Publ., St. Paul, MN, pp. 265–285.

Koerner, R. M., Koerner, G. R., and Eberlé, M. A. (1996). "Out-of-Plane Tensile Behavior of Geosynthetic Clay Liners," Geosynthetics International, Vol. 3, No. 2, pp. 277–296.

Koerner, R. M., Koerner, G. R., and Hwu, B.-L. (1990). "Three Dimensional, Axi-Symmetric Geomembrane Tension Test," Geosynthetic Testing for Waste Containment Applications, ASTM STP 1081, Robert M. Koerner, Ed., American Society for Testing and Materials, Philadelphia, PA, pp. 170–184.

Koerner, R. M., and Narejo, D. (1995). "Bearing Capacity of Hydrated Geosynthetic Clay Liners," Tech. Note Jour. of Geotechnical Engineering Division, ASCE, Vol. 121, No. 1, pp. 82–85.

Lee, C. R., Skogerboc, J. G., Eskew, K., Price, R. W., Page, N. R., Clar, M., Kort, R., and Hopkins, H. (1984). "Restoration of Problem Soil Material at Corps of Engineers Construction Sites," Instruction Report EL-84-1, U.S. Army Engineer Waterways Experiment Station, Vicksburg, MS.

Ligotke, M. S., and Klopfer, D. C. (1990). "Soil Erosion Rates from Mixed Soil and Gravel Surfaces in a Wind Tunnel," PNL-7435, Pacific Northwest Laboratory, Richland, WA.

Mayne, P. W., Jones, J. S., and Dames, J. C. (1984). "Ground Response to Dynamic Compaction," Jour. Geotechnical Engineering Div., ASCE, Vol. 110, No. 6, June, pp. 757–774.

Nyhan, J. W., Abeele, W. V., Drennon, B. J., Herrera, W. J., Lopez, E. A., Landhorst, G. J., Stallings, E. A., Walder, R. D., and Martinez, J. L. (1985). "Development of Technology for the Design of Shallow Land Burial Facilities at Arid Sites," LA-UR-35-3278, Proceedings of the Seventh Annual Participants' Information Meeting, DOE, Low-Level Waste Management Program.

Repa, E. W. J., Herrmann, E. F., Tokarski, E. F., and Eades, R. R. (1987). "Evaluating Asphalt Cap Effectiveness at a Superfund Site," Jour. Env. Eng., Vol. 113, No. 3, June, pp. 649–653.

Soong, T.-Y., and Koerner, R. M. (1996). "Seepage Induced Slope Instability," Proc. GRI-9 Conf. on Geosynthetics in Infrastructure Enhancement and Remediation, GII Publ., Philadelphia, PA, pp. 245–265.

Swope, G. L. (1975). "Revegetation of Landfill Cover Sites," M.S. Thesis, Pennsylvania State University, State Park, PA.

Thornburg, A. A. (1979). "Plant Materials for Use on Surface Mined Lands," TP-157 and EPA-600/7-79-134, Soil Conservation Service, U.S. Department of Agriculture, Washington, DC.

Trautwein, S. J., and Boutwell, G. P. (1994). "In Situ Hydraulic Conductivity Tests for Compacted Clay Liners and Caps," *Hydraulic Conductivity and Waste Contaminant Transport in Soil*, ASTM STP 1142, D. E. Daniel and S. J. Trautwein, Eds., American Society for Testing and Materials, Philadelphia, PA, pp. 184–223.

Wing, N. R., and Gee, G. W. (1994). "Quest for the Perfect Cap," Civil Engineering, 64(10), pp. 38–41.

Wischmeier, W. H., and Smith, D. D. (1960). "A Universal Soil-Loss Equation to Guide Conservation Form Planning," 7th Intl. Conf. on Soil Science, Madison, WI.

Wright, M. J. (Ed.) (1976). *Plant Adaptation to Mineral Stress in Problem Soils*, Cornell University Agricultural Experiment Station, Ithaca, NY.

CHAPTER 3

FINAL COVER SYSTEM CROSS SECTIONS

This chapter presents a number of alternative examples of cross sections of final cover systems. The chapter will be subdivided into covers for municipal solid waste (MSW), hazardous waste, and abandoned dumps. Site remediation covers are included in the category of abandoned dump covers. The emphasis throughout the chapter will be on final, rather than temporary, covers. While specific materials will be presented, it should be noted that these are not the only options that are available to the designer. They are, however, recommendations of the authors based on available materials and what is felt to be best available technology. It should be emphasized that the cross sections are presented as examples, not as exclusive recommendations. Site-specific and waste-specific conditions will always dictate the final configuration and materials.

3.1 MUNICIPAL SOLID WASTE (MSW) COVERS
Within the context of the individual components of final cover systems, namely;

- the surface layer,
- the protection layer,
- the drainage layer,
- the barrier (hydraulic and gas) layer, and
- the gas collection/foundation layer (considered as a combined layer),

there are a number of site-specific and project-specific conditions which must be addressed. With respect to MSW, the issue of allowable percolation passing the hydraulic barrier layer and into the underlying waste should be considered. Admittedly subjective, we will consider the following three categories for MSW final covers:

- minimum allowable percolation, i.e., the lowest possible amount of water being allowed to pass through the hydraulic barrier layer and into the underlying waste,
- medium allowable percolation, i.e., an intermediate amount of water being allowed to pass through the hydraulic barrier layer and into the waste, and
- maximum allowable percolation, i.e., the highest amount of water being allowed to pass through the hydraulic barrier layer (or final cover systems as a whole, if no hydraulic barrier layer is present) and into the waste.

Figures 3.1, 3.2 and 3.3 illustrate typical recommended profiles for minimum, medium, and maximum allowable percolation respectively. Furthermore, recommendations are given for humid versus arid sites. Climatological conditions at the site are very important. A humid site will generally require a different final cover system strategy than will an arid site. The potential components are compared and contrasted in Table 3.1.

Regarding the *surface layer*, the type of vegetation at humid sites should be indigenous to the local area. The topsoil thickness should be based upon the depth of the root system of the particular plant species being utilized. For arid sites, the size of cobbles depends on local conditions, particularly wind velocity, and the geometric configuration of the cover. Other hard armor systems are also available as an alternative to cobbles. Grout-filled mattresses and articulating precast concrete blocks are examples of alternative materials (Koerner 1994).

Regarding the *protection layer*, a number of issues should be considered, thickness being the major one. Of the various reasons for protection layers mentioned in Chapter 2, frost protection is always raised when the site is located in areas with significant depth of frost penetration. However, frost penetration is a critical issue only when a CCL is used in the cover system. As described by Othman et al. (1994), CCLs generally lose their low hydraulic conductivity when exposed to even one freeze-thaw cycle, let alone many cycles. Neither geomembranes nor GCLs are sensitive to freeze-thaw cycling, e.g., see Comer et al. (1995) and Hewitt and Daniel (1997) respectively. Thus, protection layer thicknesses of many meters are neither needed nor desired when using geomembranes and/or GCLs. The thickness of the protection layer in the absence of frost penetration concerns is dictated by issues such as human intrusion, burrowing animals, root penetration, potential for desiccation, etc.

Regarding the *drainage layer*, the climatology of the site is a key issue. As mentioned in Chapter 1, the drainage layer can serve several purposes, but enhancement of slope stability is typically the most important. Inadequate drainage has created slope stability problems in the past [see Boschuk

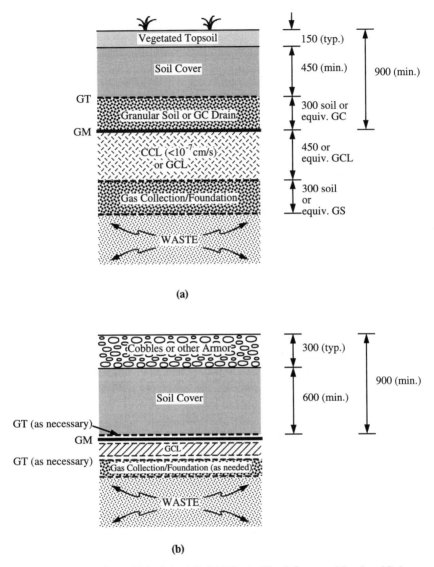

(a)

(b)

FIG. 3.1. *Examples of Municipal Solid Waste Final Covers Allowing Minimum Percolation (All Values in mm, Unless Noted)*

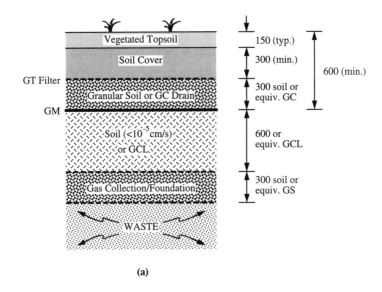

(a)

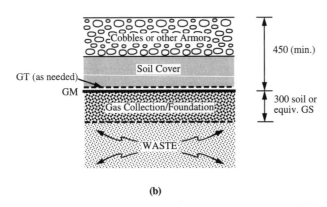

(b)

FIG. 3.2. *Examples of Municipal Solid Waste Final Covers Allowing Medium Percolation (All Values in mm, Unless Noted)*

(1991), Thiel and Steward (1993), and Soong and Koerner (1996)]. Chapter 5 will focus on slope stability methods and related details.

Regarding the *hydraulic/gas barrier layer*, a composite GM/CCL or GM/GCL is the most effective barrier system available and provides for the minimum percolation. A geomembrane or GCL alone is slightly less

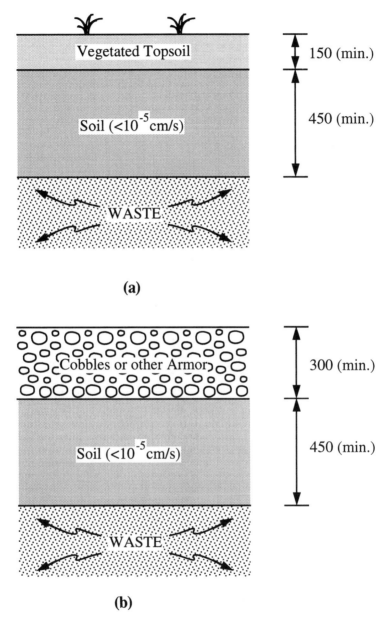

(a)

(b)

FIG. 3.3.   Examples of Municipal Solid Waste Final Covers Allowing
Maximum Percolation (All Values in mm, Unless Noted)

TABLE 3.1. Comparison of Components for Municipal Solid Waste Landfill Final Covers

Component	Minimum Percolation		Medium Percolation		Maximum Percolation	
	Humid	Arid	Humid	Arid	Humid	Arid
Surface layer	Vegetated topsoil	Cobbles	Vegetated topsoil	Cobbles	Vegetated topsoil	Cobbles
Minimum soil protection layer thickness including surface layer (in mm)	900*	900	600*	450	600*	750
Drainage layer	Soil or GC	None	Soil or GC	None	None**	None
Barrier layer (hydraulic and gas control)	GM/CCL or GM/GCL	GM/GCL	GM/CCL or GM/GCL	GM	Low permeability soil	Low permeability soil
Gas collection layer	Soil or GS	Soil or GS	Soil or GS	Soil or GS	Soil or GS	Soil or GS
Foundation layer	As needed	As needed	As needed	As needed	As needed	As needed

Notes: GC = geocomposite   CCL = compacted clay liner
GM = geomembrane   GCL = geosynthetic clay liner
GS = geosynthetic
*Greater if needed to protect drainage layer or barrier layer from freezing.
**Unless required for slope stability.

efficient and provides for medium percolation. A low permeability soil, e.g., initially at $k \leq 10^{-7}$ cm/s, used alone in the hydraulic barrier layer will deteriorate over time and should only be used when the maximum percolation is acceptable. The amount of percolation that is acceptable is a waste-specific and site-specific decision. Allowable percolation also interrelates with the liquid management strategy practiced at the site, as discussed in Chapter 1. The choice of type of geomembrane will often be controlled by the need for relatively high out-of-plane, or axisymmetric, deformability. Different types of geomembranes have been evaluated and have been the subject of a number of studies [see Steffen (1986), Koerner et al. (1990), and Nobert (1993)]. The results of these studies usually lead to the use of very flexible polyethylene (VFPE), flexible polypropylene (fPP), or polyvinyl chloride (PVC). The VFPE group consists of very low density polyethylene (VLDPE), linear low density polyethylene (LLDPE), and low density linear polyethylene (LDLPE). Alternatively, coextruded polyethylenes, like HDPE/VFPE/HDPE, are also available. The choice of the underlying soil material, CCL or GCL, has been discussed in Chapter 2, and is controlled primarily by issues related to differential settlement, freeze/thaw cycling, wet/dry potential, and slope stability.

Regarding the need for a *gas collection layer* located beneath the barrier material, one is strongly recommended for MSW when a geomembrane is in the barrier layer or is a component of a composite barrier layer. Very large geomembrane "blow-outs" caused by trapped gas beneath the geomembrane have occurred at a number of landfill final covers (recall Figure 1.7b). Even if there is not a blow-out, gas pressures transmitted to the base of the geomembrane can lower the normal stress at the GM/GCL or GM/CCL interface, thereby creating slope instability. The material for the gas collection layer will be either sand, a drainage geocomposite (e.g., geotextile/geonet/geotextile or geotextile/drainage core/geotextile), or a thick needle-punched nonwoven geotextile. Only in the case of a barrier layer allowing for maximum percolation, consisting of a relatively high permeability soil as the overlying barrier layer, can a gas collection layer possibly be omitted. Even then, the gas collection layer may still be desirable if maximum control of gas emission is desired.

Regarding a *foundation layer*, the existing surface of the MSW should be proof-rolled with compactors to avoid settlement/subsidence as much as possible. In the extreme, the waste mass can be densified using deep dynamic compaction (see Galenti et al. 1991). Upon suitable compaction, the waste surface will generally have a soil foundation layer that is graded according to the plans and specifications. This foundation layer can be a granular soil, which acts as the gas collection layer, or a fine-grained soil that would then require a geosynthetic gas collection layer placed above.

3.2 HAZARDOUS WASTE COVERS

Final covers for hazardous waste landfills consist of the same individual components as covers for municipal solid waste landfills. With hazardous waste covers, however, the general strategy for the barrier layer is to keep the underlying waste as dry as possible. Thus, minimum allowable percolation, through a composite barrier consisting of a GM/CCL or GM/GCL, is almost always recommended. The recommended cross sections are different for humid versus arid climates. Additionally, the possibility of using a lightweight fill to minimize post-construction settlement/subsidence should be considered for compressible wastes or underlying subgrade soils, for example, in closing ponds and lagoons containing highly compressible sludges. Thus, the following categories will be considered in this section:

- humid areas,
- arid areas, and
- lightweight fills.

Figures 3.4, 3.5 and 3.6, respectively, illustrate examples of cross sections for these three situations. The contrasting components can be further examined in Table 3.2.

Regarding the *surface layer*, the same comments stated previously for MSW final covers apply here as well.

Regarding the *protection layer*, thickness is again the issue, but not only from a freezing perspective but also from the point of view of ensuring physical separation between the underlying waste and potential human intruders, burrowing animals, or plant roots. For radioactive, combustible, or toxic wastes, the thickness of the soil cover will often be greater than the values shown in the examples presented in Figures 3.4, 3.5 and 3.6 and listed in Table 3.2 to provide greater protection for design service periods of hundreds of years or more. The thickness of the protection layers for these special classes of waste is a waste-specific and site-specific issue.

Regarding the *drainage layer*, the local climate is the governing consideration. Surface water migrating through the cover soil can be accommodated by a natural soil drainage material (sand or gravel) or by a drainage geocomposite placed above the barrier layer. While such drainage is within the state-of-the-practice, an inadequate drainage system has been the cause of final cover soil instability in a number of slope failures (Boschuk 1991).

Regarding the *hydraulic/gas barrier layer*, a composite system is almost always recommended for hazardous waste. The composite barrier system will typically be a geomembrane placed over a CCL or GCL. For hazardous wastes in humid areas, as illustrated in Figure 3.4, the use of a CCL is often appropriate because total and differential subsidence of hazardous wastes is often small and the incidence of cracking is less than in MSW landfills or abandoned dumps.

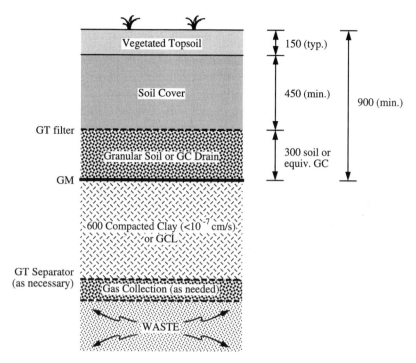

FIG. 3.4.   Examples of Hazardous Waste Final Cover in Humid Area (All Values in mm, Unless Noted)

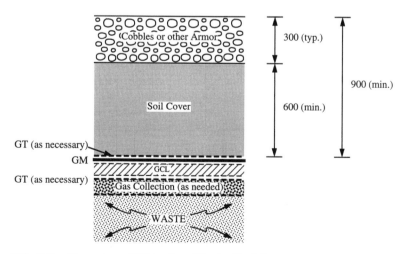

FIG. 3.5.   Examples of Hazardous Waste Final Cover in Arid Area (All Values in mm, Unless Noted)

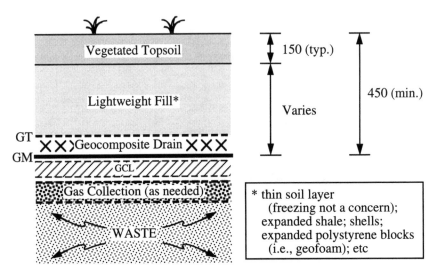

FIG. 3.6.   Examples of Lightweight Hazardous Waste Final Cover in Humid Area
(All Values in mm, Unless Noted)

TABLE 3.2.   Comparison of Components for Hazardous Waste Landfill Final Covers

Component	Humid Area	Arid Area	Lightweight Cover (in Humid Area)
Surface layer	Vegetated topsoil	Cobbles	Vegetated topsoil
Minimum soil protection layer thickness including surface layer (in mm)	900*	900	450*
Drainage layer	Soil or GC	None	GC
Barrier layer (hydraulic and/or gas control)	GM/CCL or GM/GCL	GM/GCL	GM/GCL
Gas collection layer	As needed	As needed	As needed
Foundation layer	As needed	As needed	As needed

Notes: GC = geocomposite   CCL = compacted clay liner
         GM = geomembrane   GCL = geosynthetic clay liner
*Greater if needed to protect drainage layer or barrier layer from freezing.

For abandoned dumps, closed sludge ponds, and certain other sites containing hazardous wastes, where large settlement is expected, a GCL may prove to be a better choice than a CCL. The ultimate decision of using a CCL or GCL will depend on other project-specific factors, technical issues, and cost. For the arid areas and lightweight fills shown in Figures 3.5 and 3.6, the barrier system will usually be a GM/GCL composite.

Regarding a *gas collection layer* under the barrier layer, the nature of the underlying waste must be considered. If there is no likelihood of rising gases, the use of a gas collection layer is not necessary.

Regarding a *foundation layer* upon which the final cover system is constructed, its need is a waste-specific consideration. Many hazardous waste landfills have large thicknesses of soil between the waste itself (which is often in horizontal layers, in trenches, or compartmentalized) and the ground surface. In such cases, the waste mass is relatively stable in comparison to MSW landfills and abandoned dumps. Thus, the final cover layer placed above the last lift of waste can be used as a foundation layer after it is properly graded and compacted.

3.3 ABANDONED DUMPS AND WASTE REMEDIATION COVERS

Final covers for abandoned dumps and remediation sites consist of the same individual components as covers for other types of waste.

In contrast with MSW and hazardous waste sites, relatively little is generally known about the properties and variability of the waste mass. Even the depth and lateral extent of the waste mass may be in question, since exploratory drilling can pose safety issues and related concerns. Thus, a relatively conservative approach toward final cover design is recommended. We will consider strategies of minimum and medium allowable percolation, similar to MSW, and also a special class of steep slopes. While this latter category could also pertain to MSW and hazardous waste final covers, the typical designed and regulated nature of MSW and hazardous waste sites will usually give rise to final grades that are stable or can be made stable, e.g., by the use of veneer reinforcement. Often 3(H)-to-1(V), i.e., 18.4 deg., is the maximum allowable slope for MSW and hazardous waste final covers, and in some state regulations it is only 4(H)-to-1(V), i.e., 14.0 deg.

For abandoned dumps and waste remediation projects, restricting final covers to gentle slopes is often not possible. Furthermore, site grading and lowering of existing slopes may not be possible due to safety or emission concerns. Thus, slopes of 2(H)-to-1(V), i.e., 26.6 deg. and higher, are sometimes encountered. Such slopes usually call for some type of geosynthetic reinforcement, and in almost all cases, slope stability (rather than water percolation) should be the principal design concern. The most effective system for percolation control is useless if the cover slides off of the slope.

The cross sections of Figure 3.7, 3.8, and 3.9 illustrate examples of cross sections that will be considered in this section. The examples provide the following:

- minimum percolation,
- medium percolation, and
- steep slopes, respectively.

All are illustrated for humid and arid climates. Table 3.3 counterpoints the basic components of the different examples.

Regarding the *surface layer*, the same comments as in the MSW section apply.

Regarding the *protection layer*, the situation is also similar to that of MSW, except for the steep slope cross sections shown in Figure 3.9. Depending on the slope angle, the distance between benches, the soil type, and eventual use of the site (if any), the design may call for soil reinforcement. This reinforcement is called "veneer reinforcement" and can be provided by geogrids or high-strength geotextiles. It is usually placed within the protection layer of a final cover cross section. This topic will be discussed further in Chapter 5.

Regarding the *drainage layer*, abandoned dump and remediation sites located in humid areas will generally require such a layer. Arid sites are often designed without such a provision. If required, drainage can be provided by a soil layer (sand or gravel) or by a geosynthetic drainage material. The latter will be a drainage core with a geotextile filter/separator on the surfaces, a geonet with a geotextile filter/separator on the surfaces, or a thick needle-punched nonwoven geotextile. The decision should be based on the required flow rate, as determined by procedures that were outlined in Chapter 2.

Regarding the *hydraulic/gas barrier layer*, a composite system is usually recommended. This consists of a geomembrane over a CCL or GCL. Unless the abandoned waste is very well characterized and known, however, the choice will generally be a GM/GCL. GCLs can sustain a reasonable amount of differential settlement [see Daniel and Boardman (1994) and Koerner et al. (1996)]. Conversely, CCLs are particularly vulnerable to damage from differential subsidence. The steep slope situation of Figure 3.9 is extremely site-specific. Slopes at some abandoned waste sites can be so steep as to even challenge the need for a barrier layer at all.

Regarding the *gas collection layer*, one is recommended because the existence of gas is generally a possibility. The layer itself can be natural soil (sand or gravel), or a geosynthetic. The geosynthetic will usually be a geonet with a geotextile filter/separator on the surfaces or a thick needle-punched nonwoven geotextile. The key design issue is an estimate of the

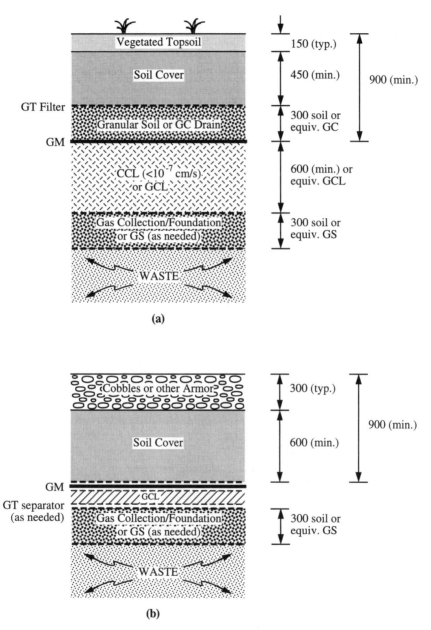

FIG. 3.7. Examples of Abandoned Dump or Remediation Site Final Covers Allowing Minimum Percolation (All Values in mm, Unless Noted)

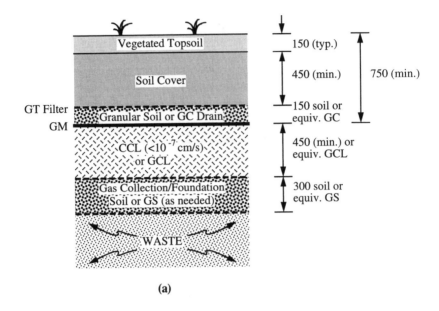

(a)

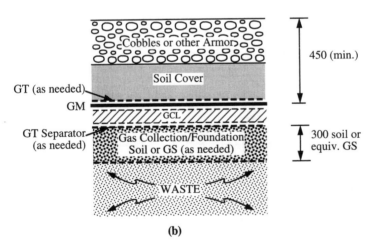

(b)

FIG. 3.8.   Examples of Abandoned Dump or Remediation Site Final Covers Allowing Medium Percolation (All Values in mm, Unless Noted)

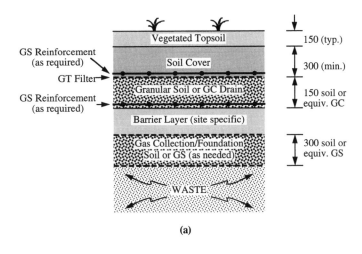

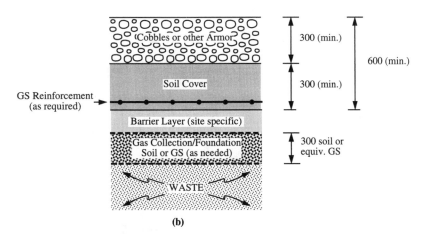

FIG. 3.9. *Examples of Abandoned Dump or Remediation Site Final Covers on Steep Slopes Greater than 3(H)-to-1(V), i.e., 18.4° (All Values in mm, Unless Noted)*

gas generation rate, for which there is no available literature to our knowledge. Thus, a conservative estimate of gas flow rate capacity is warranted.

Regarding a *foundation layer*, the abandoned waste must be heavily proof-rolled. It is best to begin with a soil layer, if one is not already present at the site. Deep dynamic compaction has been performed on abandoned waste sites, but worker's safety concerns must be considered. The foundation layer could be a granular soil and thus serve as the gas collection

TABLE 3.3. Comparison of Components for Abandoned Waste Landfill Final Covers

Concept	Minimum Percolation		Medium Percolation		Steep Slopes	
	Humid	Arid	Humid	Arid	Humid	Arid
Surface layer	Vegetated topsoil	Cobbles	Vegetated topsoil	Cobbles	Vegetated topsoil	Cobbles
Minimum soil protection layer thickness including surface layer (in mm)	900*	900	750*	450	600*	600
Drainage layer	Soil or GC	None	Soil or GC	None	Soil or GC	None
Barrier layer (hydraulic and gas control)	GM/CCL or GM/GCL	GM/GCL	GM/CCL or GM/GCL	GM/GCL	Site-specific	Site-specific
Gas collection layer	Soil or GC (as needed)	Soil or GC (as needed)	Soil or GC (as needed)	Soil or GC (as needed)	Soil or GC (as needed)	Soil or GC (as needed)
Foundation layer	As needed	As needed	As needed	As needed	As needed	As needed

Notes: GC = geocomposite   CCL = compacted clay liner
          GM = geomembrane   GCL = geosynthetic clay liner
*Greater if needed to protect drainage layer or barrier layer from freezing.

layer as well. Lastly, it should be mentioned that the foundation layer is the soil material from which final grade is generally made. It must be placed according to the final site-specific plans and specifications. After its placement, compaction, and grading, uniform thickness soil layers and various geosynthetics are placed above. If the grades in the foundation layer are inaccurate, every material that follows will be affected accordingly. All final covers are similar in this regard.

3.4 REFERENCES

Boschuk, J. J. (1991). "Landfill Covers: An Engineering Perspective," Geotechnical Fabrics Report, Vol. 9, No. 2, March, IFAI, pp. 23–34.

Comer, A. I., Sculli, M. L., and Hsuan Y, G. (1995). "Effects of Freeze-Thaw Cycling on Geomembrane Sheets and Their Seams," Proc. Geosynthetics '95, IFAI, St. Paul, MN, pp. 853–866.

Daniel, D. E., and Boardman, B. T. (1993). "Report of Workshop on Geosynthetic Clay Liners," U.S. Environmental Protection Agency, Risk Reduction Engineering Laboratory, Cincinnati, Ohio, EPA 600/2-93/171.

Galenti, V. N., Eith, A. E., Leonard, M. S. M., and Feen, P. S. (1991). "An Assessment of Deep Dynamic Compaction as a Means to Increase Refuse Density for an Operating Municipal Waste Landfill," Proc. of Waste Conference by Midland Geotechnical Society, United Kingdom, July, pp. 183–193.

Hewitt, R. D. (1994). "Hydraulic Conductivity of Geosynthetic Clay Liners Subjected to Freeze/Thaw," M.S. Thesis, University of Texas at Austin, 103 pgs.

Koerner, R. M. (1994). Designing with Geosynthetics, 3rd Edition, Prentice Hall Publ. Co., Englewood Cliffs, NJ, 783 pgs.

Koerner, R. M., Koerner, G. R., and Eberlé, M. A. (1996). "Out-of-Plane Tensile Behavior of Geosynthetic Clay Liners," Geosynthetics International, Vol. 3, No. 2, pp. 277–296.

Koerner, R. M., Koerner, G. R., and Hwu, B.-L. (1990). "Three Dimensional, Axi-Symmetric Geomembrane Tension Test," Proc. Geosynthetic Testing for Waste Containment Applications, R. M. Koerner, Ed., ASTM STP 1081, West Conshohocken, PA, pp. 170–184.

Nobert, J. (1993). "The Use of Multi-Axial Burst Test to Assess the Performance of Geomembranes," Proc. Geosynthetics '93 Conference, Vancouver, Canada, IFAI Publ., pp. 685–702.

Othman, M. A., Benson, C. H., Chamberlain, E. J., and Zimmie, T. F. (1994). "Laboratory Testing to Evaluate Changes in Hydraulic Conductivity of Compacted Clays Caused by Freeze-Thaw: State-of-the-Art," in Hydraulic Conductivity and Waste Containment Transport *in Soils, ASTM STP 1142*, D. E. Daniel and S. J. Trautwein, Eds., ASTM, Philadelphia, pp. 227–254.

Soong, T.-Y., and Koerner, R. M. (1996). "Seepage Induced Slopes Instability," Jour. Geotextiles and Geomembranes, Vol. 14, Nos. 7/8, pp. 425–445.

Steffen, H. (1984). "Report on Two Dimensional Strain Stress Behavior of Geomembranes With and Without Friction," Proc. Intl. Conf. on Geomembranes, Denver, CO, USA, IFAI Publ., pp. 181–186.

Thiel, R. S., and Stewart, M. G. (1993). "Geosynthetic Landfill Cover Design Methodology and Construction Experience in the Pacific Northwest," Geosynthetic '93 Conference Proceedings, IFAI Publ., St. Paul, MN, pp. 1131–1144.

CHAPTER 4

WATER BALANCE ANALYSIS

4.1 INTRODUCTION

One of the most important functions of a landfill cover is to limit or eliminate the production of leachate in underlying waste by minimizing or eliminating percolation of water through the cover. The analysis of water routing in covers is called *water balance analysis*. The reasons why designers or regulators analyze water balance in covers may include one or more of the following:

1.   To compare alternative design profiles and materials.
2.   To help to understand how the cover will function and which water routing mechanisms are most important.
3.   To estimate flow rates so that components of the system (e.g., pipes and geosynthetics) can be sized properly.
4.   To estimate the amount of contaminated liquid that will be generated. This value can be used as input to a fate-and-transport model of impacts on ground water. Fate-and-transport modeling is often a critical component of risk-based corrective action for site remediation projects

The fourth objective listed above is probably the most underutilized one (especially for landfill covers) and is the subject of the last section of this chapter.

4.2 WATER ROUTING WITHIN COVERS

The potential pathways for water movement onto and through a cover are summarized in Figure 4.1. The input of water is precipitation, and output is drainage (*percolation*) of water out of the cover. Within the cover, water can be stored, drained laterally, or be returned to the atmosphere via evapotranspiration. To conserve mass, the quantity of water that flows into the cover must equal the quantity of flow out of the cover plus the change

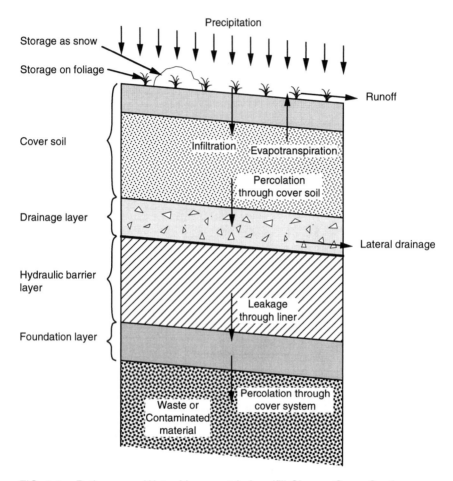

FIG. 4.1. *Pathways or Water Movement in Landfill Closure Cross Section*

in amount of water stored within the cover. This principle of conservation of mass is the basis for the term *water balance*.

Precipitation can fall on the cover in the form of either rain or snow. Most meteoric water reaches the ground surface, but some may be intercepted and stored by foliage of plants growing on the cover. Rain that reaches the surface of the cover can either run off or infiltrate the subsoil. Snow can accumulate on the surface and later melt.

Covers are usually designed to minimize the amount of percolation of water out the base of the cover. Experienced designers take advantage of all mechanisms that tend to reduce the rate of water percolation from the base of the cover. Water percolation is minimized by maximizing runoff, maximizing lateral drainage, maximizing evapotranspiration, and physically blocking downward infiltration of water by including one or more hydraulic barrier layers in the cover cross section (recall the cross sections shown in Chapter 3). The slope of the surface of the cover should be sufficiently large to accommodate both total and differential settlement and still maintain a pattern of runoff, but should not be so steep as to create stability problems or risk excessive erosion.

Water that does enter the cover flows downward by gravitational forces. However, capillary action tends to retain water in the soil. Storage of water in soil, coupled with removal of water by evapotranspiration, is the single-most important mechanism for limiting the percolation of water through the cover. Most of the water that falls on the cover infiltrates the soil, is stored in the soil, and is later returned to the atmosphere by plants through evapotranspiration.

When water infiltrates a cover, the water may be stored or it may be routed elsewhere. Near the ground surface, water can evaporate directly to the atmosphere. Within the root zone of plants, water can be withdrawn from the soil by the plants and returned to the atmosphere by the process of *transpiration*. If vertical seepage occurs within the cover, the downward-moving water can be drained laterally, if a drainage layer is present to intercept the seepage, or the water can be slowed by a hydraulic barrier layer, which is sometimes called a *resistive layer* because it provides resistance or impedance to flow.

The process of analyzing water balance in landfill covers is complex. Not only are there many possible routes that water can take, but there are also numerous complicating factors. For example, if the ground is frozen, water in the frozen zone will be unavailable for movement or rerouting. The storage of water in a frozen zone should be taken into account in a rigorous analysis, but that requires an understanding of the depth of frost penetration into soil and how the depth of frost penetration changes with time. Also, an understanding of the fate of water that reaches the cover is only as good as the knowledge of precipitation patterns and other relevant climatological factors. Because we cannot predict future precipitation and other weather-related parameters with confidence (except perhaps in a statistical sense), our ability to predict the performance of a cover will always be limited, particularly during and after extreme weather events. For a cover that is intended to allow little or no percolation of water into underlying waste or contaminated material, it may be desirable to account for these uncertainties by employing a conservative design approach and incorporating redundant design features into the cross section, e.g., by including a resistive barrier layer even though one may not be needed. However, by the same token, a layer should not arbitrarily be added to the cross section of the cover unless that layer will improve the performance of the cover.

All too often, layers are included in covers simply to comply with standard regulatory guidance rather than to serve specific and necessary functions.

4.3 PRINCIPLES OF WATER BALANCE ANALYSIS

In this section, we will examine the general principles of water balance analysis and discuss the basic methods that are commonly used.

4.3.1 Retention of Precipitation by Leaves of Plants

During the initial stages of precipitation in the form of rain, most that falls on a cover is intercepted by foliage. However, most plant leaves have very limited storage capacity, and this storage capacity is gradually exceeded by further precipitation.

Schroeder et al. (1994) recommend that the interception storage capacity of the foliage be calculated based on the density of the vegetation (units are kilograms of biomass per hectare of area), but suggest that the value be between 0 and 1.3 mm of precipitation, with 1.3 mm being a reasonable estimate of the maximum amount of precipitation that can be stored by a good stand of most types of non-woody vegetation. Schroeder et al. further recommend that the fraction of storage capacity intercepted ("F," defined as the amount of intercepted water divided by the interception storage capacity) be calculated as follows:

$$F = [1 - e^{-(R/INT)}] \qquad (4.1)$$

where R is the daily rainfall and INT is the maximum interception storage capacity. The amount of water reaching the ground is the amount of rainfall minus the amount stored by plant interception. If hourly precipitation is analyzed to account for storm events, it is recommended that retention of precipitation by leaves of plants be ignored.

4.3.2 Storage of Snow at the Surface

Snow that falls to the surface of the cover is assumed to accumulate on the surface until it: (1) evaporates due to sublimation; (2) melts due to temperature rising above freezing; or (3) melts due to rainfall. Sublimation is typically ignored because it usually plays a minor role in the water balance. Schroeder et al. (1994) provide recommended procedures for calculating snowmelt from the other two mechanisms. Direct (non-rain) snowmelt depends primarily on the surface air temperature, but is affected by the amount of runoff of melted snow (vs. storage within the snow pack) and the tendency of water from snow melt to re-freeze. Snowmelt caused by rainfall is analyzed based on a heat balance. The warmer the temperature of the rain, and the greater the amount of rain, the greater the snowmelt.

4.3.3 Runoff

The runoff coefficient, "C," is defined as the ratio of runoff to precipitation. For example, if $C = 0.1$, then 10% of the precipitation is assumed to run off and 90% is assumed to infiltrate the soil. Runoff is one of the most difficult parameters to determine accurately because very little information is available on actual runoff rates from landfill covers. Most of the information that is available comes from data collected from small watersheds in the eastern US states. Comparatively few data are available for sites in arid regions or from actual landfill covers. Furthermore, the runoff coefficient is affected by parameters such as the water content of the soil, the density of vegetation, the intensity and duration of rainfall, the type of soil, the slope inclination, and probably others as well. The assumed runoff coefficient should be viewed as an approximation.

Two approaches are used for estimating the runoff coefficient. The simplest approach is to estimate a value based on the type of soil and average angle of the slope. The guidance provided by Fenn et al. (1975), which is summarized in Table 4.1, is recommended if no better information is available for the specific site. For winter months, if precipitation occurs as snowfall, the runoff coefficient should be adjusted empirically.

The procedure recommended by Schroeder et al. (1994) is more complicated and involves use of the Soil Conservation Service (SCS) curve-number method. The method, which is applicable for large storms on small watersheds, was developed by plotting measured runoff in streams versus rainfall. The typical trend that has been observed is shown in Figure 4.2. Initially, no runoff occurs, but with time and continuing rainfall, the slope of the diagram approaches 45 degrees, indicating that all of the rainfall is running off. A series of curves were developed (hence the name, SCS curve-number method), which are used along with information on soil moisture to compute runoff. Schroeder et al. (1994) describe the procedure used in a computer program as well as a method used to correct the curve numbers for steeply sloping landfill covers.

TABLE 4.1.   Suggested Runoff Coefficients (from Fenn et al. 1975)

Description of Soil	Slope	Runoff Coefficient
Sandy Soil	Flat ($\leq 2\%$)	0.05–0.10
Sandy Soil	Average (2–7%)	0.10–0.15
Sandy Soil	Steep ($\geq 7\%$)	0.15–0.20
Clayey Soil	Flat ($\leq 2\%$)	0.13–0.17
Clayey Soil	Average (2–7%)	0.18–0.22
Clayey Soil	Steep ($\geq 7\%$)	0.25–0.35

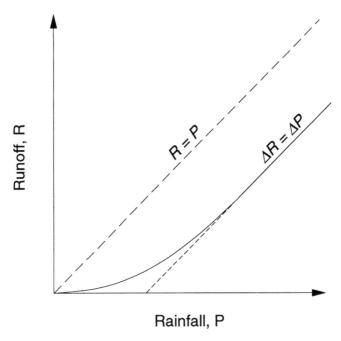

FIG. 4.2.   Typical Relationship between Runoff and Rainfall for
Large Storms in Small Watersheds

4.3.4 Storage of Water in Soil

The amount of water that a soil can store depends mainly on the type
and density of the soil, as well as the thickness of the soil layer. Figure 4.3
depicts several commonly used terms to describe soil moisture content. *Dry
soil* refers to soil that is oven dried and for all practical purposes is devoid
of water. If soils are allowed to come to equilibrium with atmospheric air
at room temperature, they will retain a small amount of moisture. (This
water content is referred to as the *hygroscopic water content.*) The hygroscopic water content of soils varies from less than 0.1% for clean gravel to
as much as approximately 10% for some extremely plastic clays, such as
bentonite.

The *wilting point* of soils refers to the water content of the soil at which
plants can no longer withdraw water and, therefore, wilt and eventually
die. Plants can typically withdraw water from soil down to water potentials
of about −15 bars (1 bar approximately equals 1 atm of pressure) before
wilting occurs. Therefore, the wilting point of a soil is typically defined as
the water content of the soil at a water potential of −15 bars. Soils at the
wilting point are obviously dry by most qualitative assessments, since they

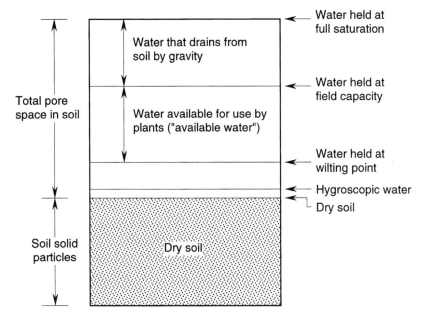

FIG. 4.3. Representation of Water Retention in Soil

will not support the growth of plants. In fact, it is frequently assumed that the wilting point is the driest that the soil is likely to become in the field.

The *field capacity* of a soil is the highest water content at which water is retained in soil without gravity drainage. When the water content of a soil rises above field capacity, water drains downward by gravity until field capacity is reached, at which point gravity drainage ceases.

The actual process of gravity drainage is fairly complex, and the measurement of the field capacity of a soil is not as simple as it might seem. The complications arise from the fact that water always flows in response to a gradient in energy, and, therefore, the water content at equilibrium is a function of the energy level in the soil water. Thus, water does not drain until a certain water content is reached. Stated differently, water drains until an equilibrium of energy in the system is reached. The energy level is typically reported as hydraulic head, or in the case of unsaturated soil, as soil suction. A rigorous analysis of the relationship between water content and drainage would require knowledge of the relationship between water content and soil suction and the variation of suction in the soil with depth and time. In fact, one standard way of measuring field capacity is to measure the relationship between water content and suction and report the field capacity as the water content at which there is small suction (typically about 10 kPa of suction). However, for the purpose of estimating water balance in landfill covers, the use of the field capacity concept expressed independently of suction does not introduce unacceptable error because the assumption regarding field capacity is but one of many approximations. Often the field capacity is estimated rather than measured, in part because the water balance is usually performed as part of the design process before the cover is constructed, when the actual cover soil material is not yet known.

Saturation is defined as the state in which all of the pore spaces of the soil are filled with water. No additional water can be retained once the soil has reached saturation.

The *available water* is the amount of water available for use by plant roots. Water is available for use by plants when the water content is above the wilting point. Because the soil is rarely wetted significantly above field capacity (and even then only for brief periods of gravity drainage), the available water is assumed to be the difference between field capacity and wilting point.

4.3.5 Evapotranspiration

Evapotranspiration refers to the transfer of water from the soil directly to the atmosphere (evaporation), and to the removal of water by plants and transfer of that water to the atmosphere by those same plants (transpiration). Of the two, the second process (transpiration) is by far the more significant because evaporation is limited to the surface of the soil, but transpiration occurs within the entire root depth of plants. The amount of evapotranspiration increases with increasing water content of the soil, surface temperature, wind speed, plant density, and depth of plant root penetration, and increases with decreasing relative humidity in the atmosphere.

The estimation of evapotranspiration is probably the most complicated part of the water balance analysis. Fenn et al. (1975) analyzed evapotranspiration using the relatively simple and largely observational-based method of Thornthwaite (Thornthwaite and Mather, 1955). Schroeder et al. (1994) used a more complex computerized method involving an energy balance (Ritchie 1972).

4.3.6 Lateral Drainage

If a lateral drainage layer is present, the maximum rate of flow within the layer is typically calculated using Darcy's formula:

$$q = k(\Delta H/L)A = k(\Delta H/L)(t)(1) \qquad (4.2)$$

where q is the rate of flow (volume/time), k is the hydraulic conductivity (length/time) of the material comprising the drainage layer, ΔH is the head loss (length) over distance L along the path of flow, and A is the cross-sectional area of flow, which is equal to the thickness of the drainage layer (*t*) times a unit distance of length. The transmissivity, *T*, of a layer is defined as the hydraulic conductivity times the saturated thickness that conducts seepage (*t*):

$$T = kt \qquad (4.3)$$

Thus, if transmissivity is used rather than hydraulic conductivity, Darcy's formula can be rewritten as:

$$q = T(\Delta H/L) \qquad (4.4)$$

Because Equation 4.4 is a simpler formulation than Equation 4.2, transmissivity is often used rather than hydraulic conductivity. Viewed in more basic terms, the flow rate is proportional to hydraulic conductivity times saturated thickness, i.e., transmissivity. The Depuit-Forcheimer assumptions are normally employed, which means that the line of seepage is assumed to be parallel to the slope of the drainage layer, and the hydraulic gradient throughout the flow regime is assumed to be constant and equal to the sine of the angle of the slope.

4.3.7 Hydraulic Barriers

Hydraulic barrier layers, discussed earlier, usually consist of one or more of the following: geomembrane, compacted clay liner (CCL), or geosynthetic clay liner (GCL). The analysis of flow through these barrier layers can be very challenging. Different types of hydraulic barrier materials are discussed in succeeding subsections.

4.3.7.1 Geomembranes. Geomembranes are non-porous materials. Water does not flow through them as it does through soil. Instead, water passes through a geomembrane by first being absorbed into the polymer structure along one face, then by diffusing through the geomembrane, and finally by desorbing from the surface at the opposite side. The proper law to simulate this process is Fick's law of diffusion, not Darcy's formula. Fick's first law of diffusion states that the rate of water movement through a material is proportional to the gradient in energy across the material. For water vapor movement, the gradient in energy is usually computed based on relative humidity differences across the geomembrane. Thus, if water is ponded on one side, the rate of diffusion across the geomembrane will increase as the relative humidity in the soil air on the opposite side decreases.

Geomembranes are difficult to install completely free of flaws. Manufacturing defects, such as pinholes, are rare but nevertheless can occur. Pinholes typically have a diameter less than the thickness of the geomembrane. Improvements in manufacturing techniques for geomembranes have virtually eliminated pinholes.

Flaws, however, can be expected from the installation of geomembranes. The flaws are the result of imperfections in the field seaming of panels, inadequate repairs, or penetrations such as those from accidental puncture. Giroud and Bonaparte (1989) evaluated flaws in geomembranes and recommended using a flaw size of 1 cm^2 for conservative (high) prediction of leakage rates. (Flaws are typically smaller than 1 cm^2.) Giroud and Bonaparte (1989) used a flaw density of 1 flaw/acre (1 flaw per 0.4 hectare) for intensely monitored projects with excellent construction quality assurance. A flaw density at least 10 times higher is probably appropriate for poorly controlled projects. There is no industry-standard practice on the assumed flaw density (since it depends on site-specific quality control and quality assurance procedures), but flaw density is usually assumed to be in the range of 1 to 5 flaws per acre (1 acre = 0.4 hectare) when there is competent construction quality assurance. Thorough manufacturing and construction quality assurance procedures have become common industry practice and are the rule rather than the exception in North America.

The flow rate through a flaw depends on the size of the flaw, head of liquid, degree of sealing of the geomembrane with the underlying soil, and hydraulic conductivity of the adjacent soils. The rate of seepage in the field will also depend on the number of penetrations through the geomembrane and the size of those penetrations. The rate of seepage will also be affected by the hydraulic conductivity of adjacent soil layers, which serve to restrict seepage in the vicinity of an imperfection in the geomembrane.

4.3.7.2 Clay Liners. Clay barrier layers in covers can consist of compacted clay liners (CCLs), geosynthetic clay liners (GCLs), or both. Water flow through both types of clay liners is analyzed using Darcy's formula:

$$q = k(\Delta H/L)A = k[(H + D)/D]A \qquad (4.5)$$

where D is the thickness of the clay liner and H is the depth of water ponded on the liner. This equation assumes that the pressure head on the base of the clay liner is zero (i.e., negligible suction exists at the clay liner/subsoil interface). In the limit, as the depth of ponded liquid (H) approaches zero, the term $(H + D)/D$, which is the hydraulic gradient, approaches unity. Gravity drainage of soil at constant water content normally occurs under unit hydraulic gradient.

Both CCLs and GCLs are installed in an unsaturated condition. A rigorous, time-dependent analysis of seepage through an initially unsaturated GCL would involve consideration of flow through an unsaturated porous medium. However, the CCL or GCL is usually assumed to be fully saturated because the hydraulic conductivity of soils increases with increasing degree of saturation. Although GCLs may initially be relatively dry, they hydrate quickly (within a few weeks) as soon as there is a source of water to cause the hydration.

4.3.7.3 Composite Liners.
A composite liner consists of a geomembrane liner placed in contact with a clay liner (either a CCL or a GCL). The analysis of seepage through composite liners is largely based upon the results of experiments and analysis of highly idealized geomembrane/clay systems. The current practice is to employ the equations published by Giroud and Bonaparte (1989) for calculating seepage rates through composite liners. The equations are presented for cases involving both good and poor hydraulic contact between the geomembrane and clay. Good contact would exist if the geomembrane and subgrade surface are both smooth so that no gaps exist between the two. Poor contact would exist if the subgrade were rough or if wrinkles existed in the geomembrane.

The equations for good and poor contact relate the quantity of seepage (q, in cubic meters per second) through a defect in the geomembrane/clay composite liner to the area (a, in square meters) of the defect in the geomembrane, the head (h, in meters) of liquid on the liner, and the hydraulic conductivity (k_s, in units of meters per second) of the clay component of the composite liner as follows:

$$q = 0.21 h^{0.9} a^{0.1} k_s^{0.74} \qquad \text{(good contact)} \qquad (4.6a)$$

$$q = 0.15 h^{0.9} a^{0.1} k_s^{0.74} \qquad \text{(poor contact)} \qquad (4.6b)$$

4.4 WATER BALANCE ANALYSIS BY HAND

Water balance analyses may be performed by hand or by a computer. In this section, the hand procedure is described. However, the procedure is highly amenable to use of a computer spread sheet to facilitate the actual computation. Three publications provide the basis of the procedures recommended here: Thornthwaite and Mather (1957), Fenn et al. (1975), and Kmet (1982).

One of the first decisions that must be made is whether to use hourly, daily, weekly, or monthly averages of precipitation. Because storm events can have a major impact on runoff, hourly averages of precipitation would be a logical time step. However, 8,760 hours comprise a year, and 8,760 time steps are too many for a hand method of analysis of an entire year. Daily averages of precipitation likewise would constitute too many time steps to be practical for analysis by hand for an entire year. Monthly averages are recommended if an entire year is to be analyzed by hand.

One of the main reasons for analyzing water balance is to determine the rate of water flow into the drainage layer (if one is present), and then to make sure that the drainage layer has adequate capacity to transmit the incoming flow. Monthly or daily averages of precipitation are useful for calculating average flows over periods of days or months, but experience indicates that during intense storms, the peak flow into drainage layers can be significantly greater than predicted from daily or monthly averages of precipitation. To maintain stable slopes, it is necessary that the slope be stable during the most critical period, which is during and immediately after intense rainstorms. Hourly precipitation data should be used to calculate peak flow rates into the drainage layer and to determine if the drainage layer has adequate capacity to transmit the peak flow.

Section 4.4.1 details the recommended procedure for analyzing the annual water balance using monthly averages of precipitation. Section 4.4.2 describes how to calculate peak water flow rates into the drainage layer for maximum storm events, based on peak hourly precipitation. Designers should consider not only peak annual flows in the drainage layer, but peak hourly flows during storm events, as well, particularly for critical slope stability situations.

4.4.1 Spread Sheet for Monthly Precipitation Data

A table or spread sheet should be set up as shown in Table 4.2. Twelve columns are established for the twelve months of the year. The following text will progress through each of the individual rows and explain the calculation procedure. An example will be presented at the end.

4.4.1.1 Row A: Average Monthly Temperature. The average monthly temperature (°C) is recorded in the first row of data (Table 4.2).

4.4.1.2 Row B: Monthly Heat Index (H_m). The monthly heat index (H_m) is a dimensionless, empirical parameter used to estimate evapotranspiration (discussed in the next section). The monthly heat index is calculated as follows:

$$H_m = (0.2T)^{1.514} \qquad \text{(for } T > 0°C) \qquad (4.7a)$$

$$H_m = 0 \qquad \text{(for } T \leq 0°C) \qquad (4.7b)$$

where T is the average monthly temperature from Row A. The monthly values are summed to determine the annual heat index (H_a), which is entered in the far right "total" column shown in Table 4.2.

4.4.1.3 Row C: Unadjusted Daily Potential Evapotranspiration (UPET). The unadjusted daily potential evapotranspiration refers to the maximum amount of evapotranspiration that would occur if the soil were saturated with water. Hence, "unadjusted" is used because the amount of actual evapotranspiration will depend on the water content of the soil and will usually be less than the unadjusted daily potential evapotranspiration.

TABLE 4.2. Spread Sheet Used for Water Balance Analysis

Row	Parameter	Reference	January	February	March	April	May	June	July	August	September	October	November	December	Total
A	Avg. Monthly Temp (°C)	Input Data													
B	Monthly Heat Index (H_m)	Eq. 4.7													
C	Unadjusted Daily Potential Evapotranspiration (UPET), mm	Eqs. 4.8 and 4.9													
D	Possible Monthly Duration of Sunlight (N)	Table 4.3 or 4.4													
E	Potential Evapotranspiration (PET), mm	PET = UPET · N													
F	Precipitation (P), mm	Input Data													
G	Runoff Coefficient (C)	See Table 4.1													
H	Runoff (R), m	$R = P \cdot C$													
I	Infiltration (IN), mm	$IN = P - R$													
J	IN − PET, mm														
K	Accumulated Water Loss (WL), mm	WL = ΣNeg. IN − PET's													
L	Water Stored (WS), mm	See 4.3.1.12													
M	Change in Water Storage (CWS), mm	See 4.3.1.13													
N	Actual Evapotranspiration (AET), in.	Eq. 4.16													
O	Percolation (PERC), mm	Eq. 4.18													
P	Check (CK), mm	Eq. 4.19													
Q	Percolation Rate (FLUX), m/s	Eq. 4.20													

The units of evapotranspiration are length (e.g., mm) by analogy to precipitation, which has the same units, e.g., mm of precipitation.

The procedure for analyzing evapotranspiration was published by Thornthwaite and Mather (1957) and is based on a procedure developed in the 1940's and 1950's by the Laboratory of Climatology in Centerton, New Jersey. Thornthwaite and his associates constructed several lysimeters at various locations around the world. A lysimeter is a large container, open at the top but sealed on the sides and base, that is embedded in the ground. The lysimeter is filled with the desired soil, and the ground surface inside the lysimeter is at the same elevation as outside. The lysimeter allows for evaluation of water balance in soil via construction of a control volume that is sealed on the sides and base. The potential evapotranspiration was found to be a function of temperature for each site, but a different function of temperature at each site because the characteristics of the sites varied. Thornthwaite plotted and analyzed the data, and found that despite the many complicating factors, the site-specific potential evaporation could be related to a parameter that he called the annual heat index (H_a, which is the sum of values from Row B, Table 4.2). The procedure recommended by Thornthwaite for computing the unadjusted daily potential evaporation (UPET) in units of millimeters is as follows:

$$\text{UPET} = 0 \qquad\qquad\qquad\qquad\qquad \text{(for } T \leq 0°C) \qquad (4.8a)$$

$$\text{UPET} = 0.53(10T/H_a)^a \qquad\qquad \text{(for } 0°C < T < 27°C) \quad (4.8b)$$

$$\text{UPET} = -0.015T^2 + 1.093T - 14.208 \quad \text{(for } T \geq 27°C) \qquad (4.8c)$$

where T is the temperature in °C, H_a is the dimensionless annual heat index, a is a dimensionless empirical factor that is computed as follows:

$$a = (6.75 \times 10^{-7})H_a^3 - (7.71 \times 10^{-5})H_a^2 + 0.01792H_a + 0.49239 \quad (4.9)$$

4.4.1.4 Row D: Monthly Duration of Sunlight (N). The mean possible monthly duration of sunlight (N), corrected for possible amount of sunlight, and expressed in units of 12 hour periods, is determined from Table 4.3 (northern hemisphere) or Table 4.4 (southern hemisphere). The value of N depends on the latitude and month of the year. The units are unimportant because this number is only used in an empirical equation later in the calculation process.

4.4.1.5 Row E: Potential Evapotranspiration (PET). The potential evapotranspiration is calculated from multiplying the values in Rows C and D.

4.4.1.6 Row F: Precipitation (P). The mean monthly precipitation (P) for the site is entered in Row F. If data are not available for the site (and

TABLE 4.3. Mean Possible Monthly Duration of Sunlight in the Northern Hemisphere
Expressed in Units of 12 Hours (from Thornthwaite and Mather, 1957)

"Northern Latitude" (deg.)	J	F	M	A	M	J	J	A	S	O	N	D
0	31.2	28.2	31.2	30.3	31.2	30.3	31.2	31.2	30.3	31.2	30.3	31.2
1	31.2	28.2	31.2	30.3	31.2	30.3	31.2	31.2	30.3	31.2	30.3	31.2
2	31.2	28.2	31.2	30.3	31.5	30.6	31.2	31.2	30.3	31.2	30.0	30.9
3	30.9	28.2	30.9	30.3	31.5	30.6	31.2	31.2	30.3	31.2	30.0	30.9
4	30.9	27.9	30.9	30.6	31.8	30.9	31.5	31.5	30.3	30.9	30.0	30.6
5	30.6	27.9	30.9	30.6	31.8	30.9	31.8	31.5	30.3	30.9	29.7	30.6
6	30.6	27.9	30.9	30.6	31.8	31.2	31.8	31.5	30.3	30.9	29.7	30.3
7	30.3	27.6	30.9	30.6	32.1	31.2	32.1	31.8	30.3	30.9	29.7	30.3
8	30.3	27.6	30.9	30.9	32.1	31.5	32.1	31.8	30.6	30.6	29.4	30.0
9	30.0	27.6	30.9	30.9	32.4	31.5	32.4	31.8	30.6	30.6	29.4	30.0
10	30.0	27.3	30.9	30.9	32.4	31.8	32.4	32.1	30.6	30.6	29.4	29.7
11	29.7	27.3	30.9	30.9	32.7	31.8	32.7	32.1	30.6	30.6	29.1	29.7
12	29.7	27.3	30.9	31.2	32.7	32.1	33.0	32.1	30.6	30.3	29.1	29.4
13	29.4	27.3	30.9	31.2	33.0	32.1	33.0	32.4	30.6	30.3	28.8	29.4
14	29.4	27.3	30.9	31.2	33.0	32.4	33.3	32.4	30.6	30.3	28.8	29.1
15	29.1	27.3	30.9	31.2	33.3	32.4	33.6	32.4	30.6	30.3	28.5	29.1
16	29.1	27.3	30.9	31.2	33.3	32.7	33.6	32.7	30.6	30.3	28.5	28.8
17	28.8	27.3	30.9	31.5	33.6	32.7	33.9	32.7	30.6	30.0	28.2	28.8
18	28.8	27.0	30.9	31.5	33.6	33.0	33.9	33.0	30.6	30.0	28.2	28.5
19	28.5	27.0	30.9	31.5	33.9	33.0	34.2	33.0	30.6	30.0	27.9	28.5
20	28.5	27.0	30.9	31.5	33.9	33.3	34.2	33.3	30.6	30.0	27.9	28.2
21	28.2	27.0	30.9	31.5	33.9	33.3	34.5	33.3	30.6	30.0	27.6	28.2
22	28.2	26.7	30.9	31.8	34.2	33.6	34.5	33.3	30.6	29.7	27.6	27.9
23	27.9	26.7	30.9	31.8	34.2	33.9	34.8	33.6	30.6	29.7	27.6	27.6
24	27.9	26.7	30.9	31.8	34.5	34.2	34.8	33.6	30.6	29.7	27.3	27.6
25	27.9	26.7	30.9	31.8	34.5	34.2	35.1	33.6	30.6	29.7	27.3	27.3
26	27.6	26.4	30.9	32.1	34.8	34.5	35.1	33.6	30.6	29.7	27.3	27.3
27	27.6	26.4	30.9	32.1	34.8	34.5	35.4	33.9	30.6	29.7	27.0	27.0
28	27.3	26.4	30.9	32.1	35.1	34.8	35.4	33.9	30.9	29.4	27.0	27.0
29	27.3	26.1	30.9	32.1	35.1	34.8	35.7	33.9	30.9	29.4	26.7	26.7
30	27.0	26.1	30.9	32.4	35.4	35.1	36.0	34.2	30.9	29.4	26.7	26.4
31	27.0	26.1	30.9	32.4	35.4	35.1	36.0	34.2	30.9	29.4	26.4	26.4
32	26.7	25.8	30.9	32.4	35.7	35.4	36.3	34.5	30.9	29.4	26.4	26.1
33	26.4	25.8	30.9	32.7	35.7	35.7	36.3	34.5	30.9	29.1	26.1	25.8
34	26.4	25.8	30.9	32.7	36.0	36.0	36.6	34.8	30.9	29.1	26.1	25.8
35	26.1	25.5	30.9	32.7	36.3	36.3	36.9	34.8	30.9	29.1	25.8	25.5
36	26.1	25.5	30.9	33.0	36.3	36.6	37.2	34.8	30.9	29.1	25.8	25.2
37	25.8	25.5	30.9	33.0	36.6	36.9	37.5	35.1	30.9	29.1	25.5	24.9
38	25.5	25.2	30.9	33.0	36.9	37.2	37.5	35.1	31.2	28.8	25.2	24.9
39	25.5	25.2	30.9	33.3	36.9	37.2	37.8	35.4	31.2	28.8	25.2	24.6

TABLE 4.3. Continued

| "Northern Latitude" | | | | | | | | | | | |
(deg.)	J	F	M	A	M	J	J	A	S	O	N	D
40	25.2	21.9	30.9	33.3	37.2	37.5	38.1	35.4	31.2	28.8	24.9	24.3
41	24.9	24.9	30.9	33.3	37.5	37.8	38.1	35.7	31.2	28.8	24.6	24.0
42	24.6	24.6	30.9	33.6	37.8	38.1	38.4	35.7	31.2	28.5	24.6	23.7
42	24.3	24.6	30.6	33.6	37.8	38.4	38.7	36.0	31.2	28.5	24.3	23.1
44	24.3	24.3	30.6	33.6	38.1	38.7	39.0	36.0	31.2	28.5	24.0	22.8
45	24.0	24.3	30.6	33.9	38.4	38.7	39.3	36.3	31.2	28.2	23.7	22.5
46	23.7	24.0	30.6	33.9	38.7	39.0	39.6	36.6	31.2	28.2	23.7	22.2
47	23.1	24.0	30.6	34.2	39.0	39.6	39.9	36.6	31.5	27.9	23.4	21.9
48	22.8	23.7	30.6	34.2	39.3	39.9	40.2	36.9	31.5	27.9	23.1	21.6
49	22.5	23.7	30.6	34.5	39.6	40.2	40.5	37.2	31.5	27.6	22.8	21.3
50	22.2	23.4	30.6	34.5	39.9	40.8	41.1	37.5	31.8	27.6	22.8	21.0

often the data are not available), data from the nearest appropriate weather station is used.

Because precipitation varies from year to year, the analyst should consider the purpose of the water balance analysis when deciding on monthly precipitation values. If the goal is to estimate the maximum expected percolation through the cover, then data for an unusually wet year should be used. If an estimate of the long-term average percolation is sought, then average monthly precipitation should be used, for instance, by selecting precipitation for a typical year.

4.4.1.7 Row G: Runoff Coefficient (C).   The dimensionless runoff coefficient, C, is defined as the ratio of runoff to precipitation. The runoff coefficient can vary widely and is very difficult to predict accurately, in large part because of the dearth of data on actual runoff from landfill covers. The guidance by Fenn et al. (1975), summarized in Table 4.1, is recommended in more applicable site-specific or region-specific information. For winter months, if precipitation occurs as snowfall, the runoff coefficient should be adjusted based upon judgment.

4.4.1.8 Row H: Runoff (R).   Runoff (R) is calculated from precipitation (P) and the runoff coefficient (C):

$$R = (P)(C) \tag{4.10}$$

4.4.1.9 Row I: Infiltration (IN).   The monthly infiltration (IN), which is defined as the amount of water entering the surface of the cover, is assumed to equal precipitation minus runoff:

$$IN = P - R \tag{4.11}$$

TABLE 4.4. Mean Possible Monthly Duration of Sunlight in the Southern Hemisphere Expressed in Units of 12 Hours (from Thornthwaite and Mather, 1957)

"Southern Latitude"

(deg.)	J	F	M	A	M	J	J	A	S	O	N	D
0	31.2	28.2	31.2	30.3	31.2	30.3	31.2	31.2	30.3	31.2	30.3	31.2
1	31.2	28.2	31.2	30.3	31.2	30.3	31.2	31.2	30.3	31.2	30.3	31.2
2	31.5	28.2	31.2	30.3	30.9	30.0	31.2	31.2	30.3	31.2	30.6	31.5
3	31.5	28.5	31.2	30.0	30.9	30.0	30.9	31.2	30.0	31.2	30.6	31.5
4	31.8	28.5	31.2	30.0	30.9	29.7	30.9	30.9	30.0	31.5	30.6	31.8
5	31.8	28.5	31.2	30.0	30.6	29.7	30.6	30.9	30.0	31.5	30.9	31.8
6	31.8	28.8	31.2	30.0	30.6	29.4	30.6	30.9	30.0	31.5	30.9	32.1
7	32.1	28.8	31.2	30.0	30.6	29.4	30.3	30.6	30.0	31.5	30.9	32.4
8	32.1	28.8	31.5	29.7	30.3	29.1	30.3	30.6	30.0	31.8	31.2	32.4
9	32.4	29.1	31.5	29.7	30.3	29.1	30.0	30.6	30.0	31.8	31.2	32.7
10	32.4	29.1	31.5	29.7	30.3	28.8	30.0	30.3	30.0	31.8	31.5	33.0
11	32.7	29.1	31.5	29.7	30.0	28.8	29.7	30.3	30.0	31.8	31.5	33.0
12	32.7	29.1	31.5	29.7	30.0	28.5	29.7	30.3	30.0	31.8	31.8	33.3
13	33.0	29.4	31.5	29.4	29.7	28.5	29.4	30.0	30.0	32.1	31.8	33.3
14	33.3	29.4	31.5	29.4	29.7	28.2	29.4	30.0	30.0	32.1	32.1	33.6
15	33.6	29.4	31.5	29.4	29.4	28.2	29.1	30.0	30.0	32.1	32.1	33.6
16	33.6	29.7	31.5	29.4	29.4	27.9	29.1	30.0	30.0	32.1	32.1	33.9
17	33.9	29.7	31.5	29.4	29.1	27.9	28.8	29.7	30.0	32.1	32.4	33.9
18	33.9	29.7	31.5	29.1	29.1	27.6	28.8	29.7	30.0	32.4	32.4	34.2
19	34.2	30.0	31.5	29.1	28.8	27.6	28.5	29.7	30.0	32.4	32.7	34.2
20	34.2	30.0	31.5	29.1	28.8	27.3	28.5	29.7	30.0	32.4	32.7	34.5
21	34.5	30.0	31.5	29.1	28.6	27.3	28.2	29.7	30.0	32.4	32.7	34.5
22	34.5	30.0	31.5	29.1	28.5	27.0	28.2	29.4	30.0	32.7	33.0	34.8
23	34.8	30.3	31.5	28.8	28.5	26.7	27.9	29.4	30.0	32.7	33.0	35.1
24	35.1	30.3	31.5	28.8	28.2	26.7	27.9	29.4	30.0	32.7	33.3	35.1
25	35.1	30.3	31.5	28.8	28.2	26.4	27.9	29.4	30.0	33.0	33.3	35.4
26	35.4	30.6	31.5	28.8	28.2	26.4	27.6	29.1	30.0	33.0	33.6	35.4
27	35.4	30.6	31.5	28.8	27.9	26.1	27.6	29.1	30.0	33.3	33.6	35.7
28	35.7	30.6	31.8	28.5	27.9	25.8	27.3	29.1	30.0	33.3	33.9	36.0
29	35.7	30.9	31.8	28.5	27.6	25.8	27.3	28.8	30.0	33.3	33.9	36.0
30	36.0	30.9	31.8	28.5	27.6	25.5	27.0	28.8	30.0	33.6	34.2	36.3
31	36.3	30.9	31.8	28.5	27.3	25.2	27.0	28.8	30.0	33.6	34.5	36.6
32	36.3	30.9	31.8	28.5	27.3	25.2	26.7	28.5	30.0	33.6	34.5	36.9
33	36.6	31.2	31.8	28.2	27.0	24.9	26.4	28.5	30.0	33.9	34.8	36.9
34	36.6	31.2	31.8	28.2	27.0	24.9	26.4	28.5	30.0	33.9	34.8	37.2
35	36.9	31.2	31.8	28.2	26.7	24.6	26.1	28.2	30.0	33.9	35.1	37.5
36	37.2	31.5	31.8	28.2	26.7	24.3	25.8	28.2	30.0	34.2	35.4	37.6
37	37.5	31.5	31.8	28.2	26.4	24.0	25.5	27.9	30.0	34.2	35.7	38.1
38	37.5	31.5	32.1	27.9	26.1	24.0	25.5	27.9	30.0	34.2	35.7	38.1
39	37.8	31.8	32.1	27.9	26.1	23.7	25.2	27.9	30.0	34.5	36.0	38.4

TABLE 4.4.   Continued

| "Southern Latitude" | | | | | | | | | | | |
(deg.)	J	F	M	A	M	J	J	A	S	O	N	D
40	38.1	31.8	32.1	27.9	25.8	23.4	25.2	27.6	30.0	34.5	36.0	38.7
41	38.1	32.1	32.1	27.9	25.8	23.1	24.9	27.6	30.0	34.5	36.3	39.0
42	38.4	32.1	32.1	27.6	25.5	22.8	24.6	27.6	30.0	34.8	36.6	39.3
42	38.7	32.4	32.1	27.6	25.2	22.5	24.6	27.3	30.0	34.8	36.6	39.6
44	39.0	32.4	32.1	27.6	24.9	22.2	24.3	27.3	29.7	34.8	36.9	39.9
45	39.3	32.7	32.1	27.6	24.9	21.9	24.0	27.3	29.7	35.1	37.2	40.2
46	39.6	32.7	32.1	27.3	24.6	21.6	23.7	27.0	29.7	35.1	37.5	40.5
47	39.9	33.0	32.1	27.3	24.3	21.3	23.4	27.0	29.7	35.1	37.8	40.8
48	40.2	33.0	32.4	27.0	24.0	21.0	22.8	26.7	29.7	35.4	38.1	41.1
49	40.5	33.3	32.4	27.0	23.7	20.7	22.5	26.7	29.7	35.4	38.4	41.7
50	41.1	33.6	32.4	26.7	23.1	20.1	22.2	26.4	29.7	35.7	38.7	42.3

Empirical adjustments may be necessary to handle snowfall, if snow remains on the surface. Some snow will evaporate due to sublimation of frozen water, but this calculation procedure does not account for this evaporation mechanism because it is relatively minor compared to other water routing mechanisms in the water balance.

4.4.1.10 Row J: Infiltration Minus Potential Evapotranspiration (IN − PET). The difference between infiltration and potential evapotranspiration (IN − PET) is computed and entered in Row J. A positive number indicates potential accumulation (storage) of water in the cover soil, i.e., more infiltration than potential evapotranspiration. If infiltration minus potential evapotranspiration is negative, then the soil is drying.

4.4.1.11 Row K: Accumulated Water Loss (WL). The accumulated water loss is the sum of the negative monthly values of IN − PET since the beginning of the year. The calculation procedure is as follows. Start with January and check the numerical value in Row J. If the value is greater than or equal to zero, enter 0.0 in Row K, and go on the next month. If the value in Row J is negative, enter the negative value in Row K. For the rest of the months (February–December) the procedure for each month (column) is as follows:

> 1.   If the value of IN − PET is ≥0, then enter the value of WL from the previous month into Row K for the month being analyzed.
> 2.   If the value of IN − PET is negative, then add this negative value to the WL from the previous month and enter in Row K.

4.4.1.12 Row L: Water Stored in the Root Zone (WS). The water stored in the root zone (WS) is defined as the amount of water (in millimeters) stored in that portion of the cover soil that can be tapped by plant roots (if any are present) for evapotranspiration. The cover soil is defined as the soil from the ground surface down to the top of the drainage layer (if present).

To calculate the amount of water stored in the root zone, one must first estimate its depth. The depth of the root zone can be difficult to estimate because it depends on climatological factors, the vegetation growing on the cover, the ability of the soil to sustain growth of a good stand of vegetation, soil moisture conditions, and perhaps other factors as well.

Schroeder et al. (1994) provide guidance on the minimum and maximum evaporative depth, as shown in Figures 4.4 and 4.5. The minimum evaporative depths, shown in Figure 4.4, which range from about 200 to 450 mm, are based loosely on literature values and analyses of water movement in unsaturated, bare soils. The maximum evaporative depths, shown in Figure 4.5, range from about 0.9 to 1.5 m. It is recommended that plant specialists be consulted to assist in estimating the depth of root penetration. However, note that the depth of root penetration is the **maximum** depth tapped by plant roots, not the depth in which most of the roots are found. Also note that when the cover is first constructed, vegetation is immature

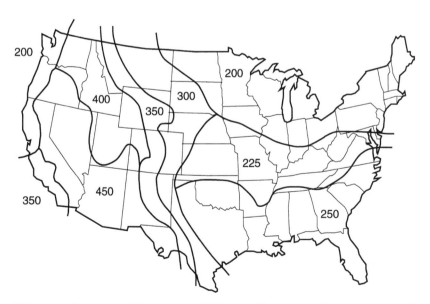

FIG. 4.4. *Geographic Distribution of Minimum Evaporative Depth (Units Are in mm)*

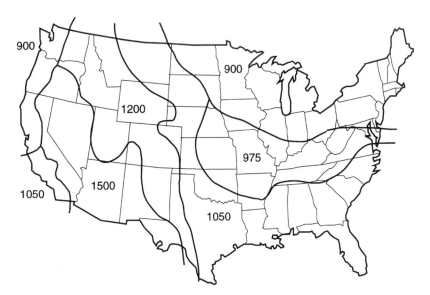

FIG. 4.5. Geographic Distribution of Maximum Evaporative Depth (Units Are in mm)

and the depth of root penetration is small. Water balance is usually analyzed for the case of mature vegetation, but it should be recognized that such conditions may not be representative of the initial conditions in the cover.

The amount of water stored in the root zone (WS) is calculated from the volumetric water content, θ, of the soil and the depth of the root zone (H_{root}) as follows:

$$WS = \theta H_{root} \tag{4.12}$$

The volumetric water content (θ) is defined as the volume of water divided by the total volume of soil (see Figure 4.6). Soil scientists and hydrogeologists are accustomed to working with water contents expressed on a volumetric basis. However, engineers typically use gravimetric water content, w, which is defined as the weight of water divided by dry weight of soil (Figure 4.6). Volumetric water content can be calculated from gravimetric water content, as shown in Figure 4.6.

Sometimes there is more than one layer of soil in the root zone, with each layer having a different volumetric water content. In such cases, the amount of stored water may be calculated by summing the amounts stored in the individual layers:

$$WS = \Sigma \theta_i (H_{root})_i \tag{4.13}$$

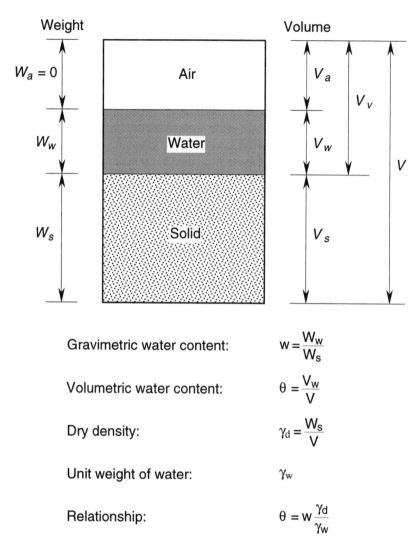

Gravimetric water content: $w = \dfrac{W_w}{W_s}$

Volumetric water content: $\theta = \dfrac{V_w}{V}$

Dry density: $\gamma_d = \dfrac{W_s}{V}$

Unit weight of water: γ_w

Relationship: $\theta = w\,\dfrac{\gamma_d}{\gamma_w}$

FIG. 4.6.   Gravimetric and Volumetric Water Content Definitions

where "i" indicates the i-th layer and the summation is performed for all the layers in the root zone.

It will be necessary to compute the maximum water storage capacity of the root zone (WS_{max}) by assuming that the entire root zone is at field capacity:

$$WS_{max} = \Sigma\theta_{i,\,\text{field capacity}}(H_{root})_i \qquad (4.14)$$

where $\theta_{i, \text{fild capacity}}$ is the volumetric water content at field capacity for the *i*-th layer. If no information is available on field capacity, the values shown in Table 4.5 are recommended. Table 4.5 also recommends values for the volumetric water content at the wilting point and the available water.

To conplete the calculation of water stored in the root zone for each month, an assumption must be made about the water content of the soil for any ore month. This selected month then becomes the starting month for the catulation process. If site-specific information is available on the actual water content in the root zone at any particular time of the year, this information may be used to estimate "θ" for a selected month. Although the water content for the starting month could be estimated without any site-specifc information, this could result in significant error and should be avoided. If the objective of the analyses is to estimate the maximum percolation rte through the cover (and this is often the objective of the water balance aialysis), the most conservative assumption to initiate the calculation proiess is that at the beginning of the wettest time of the year, the soil is at field capacity. This will ensure that some drainage is predicted.

In mos climates in North America, the winter months are the wettest, and if the cover soil is ever saturated, it is likely to be saturated in early spring. The most conservative assumption is that the cover soil is at field capacity ir an early spring month, e.g., April. There will normally be drainage from he cover soil in the ensuing (usually wet) months. However, assuming tht the soil is actually soaked to field capacity in the spring could be a bit eitreme at many sites for typical years, and a lower value should be assumed, if appropriate.

One oithe motivations for performing a water balance analysis by hand is to gain i better understanding of critical variables and assumptions. The assumption concerning the initial water content is one of the most important but least appreciated assumptions in water balance analyses. One can get almost any answer that one desires (from zero drainage of water to large drainage), depending on the assumption of initial moisture. If field capacity is assumel to be the initial moisture condition, then comparatively large

TABLE 4.5. Suggested Volumetric Water Contents of Various Soils (after Thornthwaite and Mather, 1957; and Fenn et al., 1975)

Type of Soil	θ at Field Capacity	θ at Wilting Point	Available Water
Fine Sand	0.12	0.02	0.10
Sandy Loam	0.20	0.05	0.15
Silty Loam	0.30	0.0	0.20
Clay Loam	0.375	0.125	0.25
Clay	0.45	0.15	0.30

drainage is an inevitable consequence from the calculation process. If the soil is assumed to be relatively dry initially then there may be no drainage whatsoever. One of the most important questions for reviewers of water balance analyses is to ask for assumptions about initial moisture conditions.

To compute the values of water stored in the root zone (WS) for Row L, first pick a month to start the calculation. Any month can be chosen for which the water stored is known or can be estimated. If it will be assumed that the soil is at field capacity at the end of the spring, then select the last month (usually in late spring) for which IN − PET is greater than 0.0 and assume that the water stored is equal to the water stored at field capacity, in units of mm of water stored. Enter this number in Row L.

Now proceed to the next month. The procedure to follow depends on whether IN − PET for the month of interest is positive or negative:

1. If IN − PET is negative, then the soil in the root zone will dry during the month of interest. The actual amount of soil moisture retained depends on the amount of potential evapotranspiration (i.e., the value of IN − PET) and the water storing capacity of the soil (WS_{max}). The amount of water that actually evaporates will be less than (IN − PET), and the drier the soil becomes, the more difficult it becomes to evaporate water from the soil. Compute the water stored (WS) for a given month as follows:

$$WS = (WS_{max})10^{b(IN-PET)} \qquad (4.15a)$$

where WS and WS_{max} have units of mm, b is a coefficient determined as follows:

$$b = 0.455/(WS_{max}) \qquad (4.15b)$$

and the exponent in Eq. 4.15a is b times (IN − PET), which is a negative exponent since (IN − PET) is negative. The computed value is entered into row L for the month in which IN − PET is negative.

2. If infiltration exceeds potential evapotranspiration (i.e., IN − PET > 0), add the IN − PET value for the current month to the value of water stored (WS) for the previous month. In other words, for a month in which IN − PET is positive, add the positive value of IN − PET from Row J for the month in which the calculations are being performed to the value of WS in Row L for the previous month. However, the amount of water stored cannot exceed WS_{max}, and if the computed value is greater than WS_{max}, enter the value of WS_{max} in Row L for the month.

Equations 4.15a and 4.15b were developed by fitting regression equations to data tabulated by Thornthwaite and Mather (1955).

If the soil is found to be at field capacity in the last calculation month (i.e., the month before the month used to start the calculation process), then the assumption of starting at field capacity is validated. If not, it is possible to iterate and try a different initial water storage until the computed value in the starting month is the same as the assumed value.

4.4.1.13 Row M: Change in Water Storage (CWS). Start with the same month used to initiate the process of calculating water storage in the root zone (WS) and enter 0.0 for the change in water storage (CWS) for that month. Then proceed with each subsequent month. The change in water stored in the month is the water stored (WS) in that month minus the water stored in the previous month. The sign is important: CWS is negative if the soil in the root zone is losing water and positive if it is gaining water.

4.4.1.14 Row N: Actual Evapotranspiration (AET). The actual evapotranspiration (AET) depends on whether infiltration exceeds potential evapotranspiration.

1. If $IN - PET \geq 0$, AET = PET for that month (i.e., enter the value from Row E in Row N):

$$AET = PET \qquad (\text{if } (IN - PET) \geq 0) \qquad (4.16a)$$

2. If $IN - PET < 0$:

$$AET = PET + [(IN - PET) - CWS] \quad (\text{if } (IN - PET) < 0)$$

$$(4.16b)$$

Equation 4.16 may be explained as follows. If there is less infiltration than potential evapotranspiration, the soil in the root zone is losing water through evapotranspiration. For conservation of mass, the evapotranspiration must equal infiltration minus the change in storage. Note that Equation 4.16 reduces to:

$$AET = IN - CWS \qquad (4.17)$$

4.4.1.15 Row O: Percolation (PERC). Percolation (PERC) is the amount of water draining from the root zone and is calculated as follows:

1. For months in which $IN - PET$ is less than or equal to zero, there is no percolation (evapotranspiration exceeds infiltration), and 0.0 is entered:

$$PERC = 0 \qquad (\text{if } (IN - PET) \leq 0) \qquad (4.18a)$$

2. For months in which IN − PET is greater than zero:

$$PERC = (IN - PET) - CWS \qquad (if\ (IN - PET) > 0) \qquad (4.18b)$$

The monthly percolation should be summed to obtain the annual amount of percolation.

4.4.1.16 Row P: Check of Calculations (CK). The whole idea behind "water balance" is to account for all of the precipitation that falls on the cover. The calculations may be checked as follows. For each month compute the value of CK as follows:

$$CK = PERC + AET + CWS + R \qquad (4.19)$$

and enter the value in Row P. Sum the monthly values. Each monthly value, and the yearly total, of C should equal precipitation P. Check to make sure that Row P equals Row F for each column.

4.4.1.17 Row Q: Percolation Rate (FLUX). The rate of percolation (FLUX), which is the flux of water passing through the cover soil, should be computed for months in which PERC > 0 and noted in Row Q in units of m/s. The FLUX is computed as follows:

$$FLUX = (PERC \cdot 0.001)/t \qquad (4.20)$$

where PERC is the percolation in units of mm from Row O and t is the number of seconds in the month. This information is used later (see Section 4.3.2)

4.4.1.18 Example. To illustrate the calculation process, the following example is provided based on information provided by Kmet (1982). The landfill is located in Wisconsin, which is latitude 43 degrees north. The maximum water storage in the root zone (WS_{max}) is 200 mm. The average monthly temperature and precipitation are shown in Rows A and F, respectively, of Table 4.6, along with computed values of the other parameters. The first month used to start the process of calculating water stored in the root zone is April, since April was the last month in the spring for which IN − PET was greater than zero. It was assumed that the soil was at field capacity in April and that the amount of water stored (WS in Row L) was equal to the maximum water storage in the root zone, or 200 mm. At the end of the calculation process, it was found that the soil was indeed at field capacity in April, based on conditions as they were calculated to exist in the preceding month of March.

For this example, the computed percolation over the one-year period was 72 mm. Note that all of this percolation occurred over a three month period (February, March, and April), and that there was zero percolation in all the other months.

TABLE 4.6. Example Water Balance Calculation for Landfill in Wisconsin

Row	Parameter	Reference	January	February	March	April	May	June	July	August	September	October	November	December	Total
A	Avg. Monthly Temp (°C)	Input Data	−7.1	−5.2	0.1	7.8	13.5	19.2	21.8	21.3	16.6	10.9	2.6	−4.4	
B	Monthly Heat Index (H_m)	Eq. 4.7	0.00	0.00	0.00	1.96	4.50	7.67	9.29	8.97	6.15	3.25	0.37	0.00	42.17
C	Unadjusted Daily Potential Evapotranspiration (UPET), mm	Eqs. 4.8 and 4.9	0.00	0.00	0.01	1.08	2.05	3.08	3.57	3.48	2.60	1.60	0.30	0.00	
D	Possible Monthly Duration of Sunlight (N)	Table 4.3 or 4.4	24.3	24.6	30.6	33.6	37.8	38.4	38.7	36.0	31.2	28.5	24.3	23.1	
E	Potential Evapotranspiration (PET), mm	PET = UPET · N	0.00	0.00	0.21	36.38	77.40	118.38	138.27	125.20	81.22	45.52	7.34	0.00	779.85
F	Precipitation (P), mm	Input Data	39.9	26.4	56.6	73.7	85.6	95.25	93.00	75.9	81.3	54.1	54.9	43.2	
G	Runoff Coefficient (C)	See Table 4.1	0.18	0.18	0.18	0.18	0.18	0.18	0.18	0.18	0.18	0.18	0.18	0.18	
H	Runoff (R), m	$R = P \cdot C$	7.182	4.752	10.188	13.266	15.408	17.145	16.74	13.662	14.634	9.738	9.882	7.776	140.37
I	Infiltration (IN), mm	IN = $P - R$	32.72	21.65	46.41	60.43	70.19	78.11	76.26	62.24	66.67	44.36	45.02	35.42	639.48
J	IN − PET		32.72	21.65	46.20	24.06	−7.21	−40.27	−62.01	−62.96	−14.56	−1.15	37.68	35.42	
K	Accumulated Water Loss (WL), mm	WL = ΣNeg. IN − PET's				0	−7.21	−47.48	−109.49	−172.45	−187.01	−188.16			
L	Water Stored (WS), mm	See 4.3.1.12	180.46	200.00	200.00	200.00	192.59	155.96	112.71	81.04	75.09	74.64	112.31	147.74	
M	Change in Water Storage (CWS), mm	See 4.3.1.13	32.72	19.54	0.00	0.00	−7.41	−36.63	−43.25	−31.66	−5.95	−0.45	37.68	35.42	
N	Actual Evapotranspiration (AET), in.	Eq. 4.16	0.00	0.00	0.21	36.38	77.60	114.73	119.51	93.90	72.62	44.81	7.34	0.00	567.12
O	Percolation (PERC), mm	Eq. 4.18	0.00	2.10	46.20	24.06	0.00	0.00	0.00	0.00	0.00	0.00	20.00	0.00	72.36
P	Check (CK), mm	Eq. 4.19	39.90	26.40	56.60	73.70	85.60	92.25	93.00	75.90	81.30	54.90	54.90	−43.20	779.85
Q	Percolation Rate (FLUX), m/s	Eq. 4.20	0.00E+00	8.69E−10	1.72E−08	9.28E−09	0.00E+00	0.00E+00	0.00E+00	0.00E−00	0.00E+00	0.00E+00	0.00E+00	0.00E+00	0.00E+00

TABLE 4.7. Summary of Results of Example Problem

Parameter	Annual Amount (mm)	Percent of Precipitation
Precipitation	779.85	100
Runoff	140.37	18
Actual Evapotranspiration	567.12	73
Percolation	72.36	9

Table 4.7 summarizes the annual routing of water in the cover in this example. Note that most of the water that falls on the cover infiltrates into the cover soil and is then returned to the atmosphere via evapotranspiration. This is the reason for emphasizing the proper selection of cover soil materials and plant species so that the evapotranspiration mechanism can be used to maximum advantage.

4.4.2 Analysis for Hourly Averages

Under the hypothesis that seepage induced slope instability occurs in periods consisting of hourly intervals, and recognition that the minimum time-internal from HELP is days, a manual method to calculate hourly averages is presented. Obviously, it requires hourly precipitation data. Based on the basic concepts of water balance analysis shown in Figure 4.1, the following relationships hold:

$$P = I + SR \tag{4.21}$$

and

$$I = PERC + AET + \Delta WS \tag{4.22}$$

where

P = probable maximum (hourly) precipitation (PMP)
IN = infiltration
R = surface runoff
$PERC$ = percolation
AET = actual evapotranspiration
ΔWS = change in water stored in cover soil
      = (field capacity) − (actual water content)

The conservative assumptions are made that the immediate time before the PMP event has been a period of regular rainfall, the actual evapotranspiration is negligible for an intense rainfall over a short period of time (e.g., a few hours), and the cover soil is at *field capacity* before the storm reaches its highest intensity (i.e., there is only nominal excess water storage capacity available at the time). Under these assumed conditions, the infiltration results directly in percolation, i.e., IN = PERC. Therefore, the following relationship result:

$$P = PERC + R \qquad (4.23)$$

$$\text{or } PERC = P - R$$

$$\text{but } P = P(C) \qquad (4.24)$$

where "C" equals the runoff coefficient. Thus:

$$PERC = P(1 - C) \qquad (4.25)$$

Note that Equation (4.25) is valid only when the cover soil is sufficiently permeable so that the amount of water which does not runoff [i.e., P(1 − C)] can percolate through the cover soil into the drainage layer. When the cover soil is not permeable enough to handle such amount of water, the difference will occur as sheet flow over the ground surface. The amount is governed by the hydraulic conductivity of the cover soil ($k_{cover\,soil}$). Thiel and Stewart (1993) showed that the percolation into the drainage layer, under such a situation, should be determined as;

$$PERC = k_{cover\,soil}; \qquad \text{when } P(1 - C) > k_{cover\,soil} \qquad (4.26a)$$

$$PERC = \text{as calculated} \qquad \text{when } P(1 - C) \leq k_{cover\,soil} \qquad (4.26b)$$

Equations 4.26a and 4.26b implicitly assume that the hydraulic gradient is unity. Such is the case when the head of water on the cover is small and when the water pressure at the interface between the cover soil and underlying drainage layer is equal to atmospheric pressure. On properly sloped landfill covers, it is unlikely that water could pond to significant depths on the cover, so the assumption of a small thickness of water on the surface is reasonable. However, the water pressure at the interface between the base of the cover soil and the top of the drainage layer could be slightly less than atmospheric (e.g., if the underlying drainage medium is unsaturated and there is a capillary pressure in it), or it could be greater than atmospheric if the piezometric level in the drainage layer is above the top of the drainage layer. During peak flows, with water actively percolating into the drainage layer, it is unlikely that there could be significant capillary pressure in the drainage layer. If the piezometric level in the drainage layer rises to a point above the top of the drainage layer, the hydraulic gradient would be less than unity and the percolation would be less than calculated from Eqs. 4.26a and 4.26b. However, because the purpose of this calculation is to determine the peak flow into the drainage layer, it is reasonable and usually conservative to assume that the hydraulic gradient is unity and to employ Eq. 4.26a and 4.26b.

There can be major differences in peak flows calculated using hourly, daily, and monthly averages of precipitation, as the following example of a landfill in Thrall, Texas (about 60 km northeast of Austin, Texas) illustrates. The 4-ha site has 3H:1V side slopes in both the bottom liner system and in the final cover system. For a slope length of 100 m, the calculated peak flow rates into the drainage layer were:

- Monthly precipitation: 0.0011 m^3/hr
- Daily precipitation: 0.14 m^3/hr
- Hourly precipitation: 5 m^3/hr

Note that the peak flow rate based on hourly storm data is nearly 40 times larger than the peak flow based on daily precipitation values, illustrating the importance of evaluating hourly precipitation events for purposes of calculating the adequacy of drainage layers.

4.4.3 Lateral Drainage

Water percolating from the root zone must be drained by an adequately designed underlying drainage layer. The equations for computing flow capacity in the drainage layer were presented in Section 2.3.3.

The following calculations are presented to illustrate how the required lateral drainage is analyzed. We will use the peak flow rate of 5 m^3/hr mentioned at the end of Section 4.4.2 for peak hourly precipitation.

We will assume that the landfill cover is sloped at 5% (which is 2.86°) and that the length of the slope is 100 m. Therefore, the drainage layer also has a length of 100 m. The drainage layer is assumed to have a thickness of 0.3 m and is composed of coarse sand with a hydraulic conductivity of 0.01 m/s (1 cm/s). The drainage layer is overlain by a geotextile filter, and we will assume that the filter is designed to allow water to freely discharge into the underlying drainage layer.

The flow-carrying capacity of the drainage layer is computed using Equation 4.2 and a 1-m-wide section along the slope of the cover:

$$q_{\text{flow capacity}} = k(\Delta H/L)A = (1 \text{ m/s})(\sin[2.86°])(0.3 \text{ m} \times 1 \text{ m})$$
$$= 0.015 \text{ m}^3/\text{s} \qquad (4.27)$$

The actual flow must be less than the flow-carrying capacity—otherwise the head will rise above the top of the drainage layer, possibly creating slope instability.

The peak calculated flow is 5 m^3/hr, or 0.0014 m^3/s. The rate of percolation is less than the flow-carrying capacity of the drainage layer, indicating that the drainage layer does have adequate drainage capability. To quantify this, a factor of safety (FS) may be computed as follows:

$$FS = q_{\text{flow capacity}}/q_{\text{perc}} = 0.015/0.0014 = 11 \qquad (4.28)$$

A factor of safety of at least 5 to 10 is recommended, and a factor of safety of 11 is adequate. However, note that the actual factor of safety is directly proportional to the assumed hydraulic conductivity of the drainage layer. It is often difficult to predict this hydraulic conductivity with an accuracy much better than plus or minus an order of magnitude. Conservative engineers realize this fact and often select drainage materials with significantly greater hydraulic conductivity than the minimum value assumed in design or stated in the construction specifications. If a conservative (i.e., low) estimate is made for the hydraulic conductivity of the drainage layer material, then the computed factor of safety is also conservative, and a comparatively low factor of safety calculated in Equation 4.28 (e.g., 5) might be accepted. Conversely, if a comparatively optimistic assumption is made concerning the hydraulic conductivity of the drainage layer, then a larger factor of safety would be appropriate. Conservatism is recommended because: (1) drainage layers tend to clog over time, and (2) peak daily flow rates can be higher than monthly averages.

Although the factor of safety in this example is 11, the assumed hydraulic conductivity is 0.01 m/s (1 cm/s). Many drainage layers in covers employ sand with a hydraulic conductivity of 10^{-4} m/s (0.01 cm/s). If this hydraulic conductivity had been assumed, the calculated factor of safety would have been 0.11, indicating inadequate flow carrying capacity in the drainage layer. It is the authors' opinion that if peak hourly flows are considered in design, more permeable drainage materials than those now commonly employed would be required. Because the consequences of a slope failure can be severe, the authors strongly encourage designers to consider peak hourly flows in design of drainage systems.

4.4.4 Leakage through Hydraulic Barriers

The procedures for computing leakage through hydraulic barriers were discussed in Section 4.3.7. The water that percolates through the cover soil may be drained by a drainage layer (if present) and impeded from further downward movement by one or more hydraulic barrier layers (if present).

The following calculations are presented to illustrate the process. The example problem discussed in Sections 4.4.1.18 and 4.4.2 is assumed. We will assume that the hydraulic barrier layer located beneath the drainage layer is a composite geomembrane/GCL liner. The hydraulic conductivity of the GCL is assumed to be 5×10^{-11} cm/s. The side slopes are assumed to be 300 m long. We will analyze an area of 300 m long by 10 m wide (=3,000 m^3 or 0.3 ha).

As shown in Table 4.6, for the months of February, March, and April, there was drainage from the cover soil. No drainage occurred in the other months. We will assume that leakage occurs during the three months during which water enters the drainage layer, but not during the other months.

If the geomembrane is free of defects, the leakage through the geomembrane is, for practical purposes, zero. We will assume that there is a single defect in the 0.3 ha of geomembrane, and that the area of the defect is 1 cm^2 (=0.0001 m^2). The contact between the geomembrane and bentonite is assumed to be poor since there is a likelihood that there were wrinkles in the geomembrane before it was backfilled.

The quantity of flow (Q) collected in the drainage layer during the 3 month period is:

$$Q = (72 \text{ mm} \times 0.001 \text{ m/mm})(300 \text{ m})(10 \text{ m}) = 216 \text{ m}^3 \quad (4.29)$$

and the average rate of flow (q) over the 3-month (90 d) period is:

$$q = 216 \text{ m}^3/(90 \text{ d} \times 24 \text{ h/d} \times 60 \text{ m/h} \times 60 \text{ s/m}) = 2.8 \times 10^{-5} \text{ m}^3/\text{s}$$

$$(4.30)$$

The average head of liquid in the drainage layer (h_{avg}) is conservatively computed by assuming that there is no leakage in the underlying liner:

$$q = k(\Delta H/L)A = k[\sin(2.86°)]10 \text{ m} \times h_{avg} \quad (4.31)$$

or:

$$
\begin{aligned}
h_{avg} &= q/\{k[\sin(2.86°)]10 \text{ m}\} \\
&= 2.8 \times 10^{-5} \text{ m}^3/\text{s}/\{0.01 \text{ m/s} \times \sin(2.86°) \times 10 \text{ m}\} \\
&= 5.6 \times 10^{-3} \text{ m} \quad (4.32)
\end{aligned}
$$

The leakage through the single hole assumed to exist in the geomembrane is determined from Equation 4.6b:

$$
\begin{aligned}
q &= 1.15h^{0.9}a^{0.1}k_s^{0.74} \\
&= (1.15)(5.6 \times 10^{-3} \text{ m})^{0.9}(0.0001 \text{ m}^2)^{0.1}(5 \times 10^{-11} \text{ m/s})^{0.74} \\
&= 1.0 \times 10^{-10} \text{ m}^3/\text{s} \quad (4.33)
\end{aligned}
$$

or, averaged over the 300 m $\times$ 10 m area, the flux (f) is:

$$f = q/a = 1.0 \times 10^{-10} \text{ m}^3/\text{s}/(300 \text{ m} \times 10 \text{ m}) = 3.4 \times 10^{-14} \text{ m/s} \quad (4.34)$$

In terms of amounts of liquid, recall that the amount of percolation of water from the cover soil and into the drainage layer was 72 mm per year, or 216 m^3 (Equation 4.29) for the 300 m $\times$ 10 m area being used for this example. The total flow rate from Equation 4.33, for the three month period in which there is assumed to be leakage, is 7.8×10^{-4} m^3, or approximately one millionth of the percolation entering the drainage layer. The annual leakage through the geomembrane/GCL composite barrier layer is approximately one ten-millionth of the annual precipitation. For the assumed conditions, the geomembrane/GCL barrier, for all practical purposes, stops the percolation of water through the capping system and into the underlying materials.

4.5 WATER BALANCE BY COMPUTER ANALYSIS

Nearly all water balance analyses performed for actual cover designs are conducted using the computer program HELP (Hydraulic Evaluation of Landfill Performance). Program HELP was written by Dr. P. Schroeder of the U.S. Army Corps of Engineers Waterways Experiment Station under sponsorship of the U.S. EPA. The program, which has been periodically revised, is available in the public domain. At the time of this writing, the latest version was Version 3.00 and was available by purchasing "The Hydraulic Evaluation of Landfill Performance Model, Engineering Documentation for Version 3," EPA/600/R-94/168b, from the National Technical Information Service. The user's manual is supplied with a diskette that contains the program, which was written in Fortran for use on a personal computer.

The computer program employs the same principles as the method of hand analysis described earlier, but HELP uses a daily (rather than monthly) time step and employs more sophisticated algorithms for many of the computations. The model accepts weather, soil, and design data, and uses solution techniques that account for the effects of surface storage, snowmelt, runoff, infiltration, evapotranspiration, vegetative growth, storage of soil moisture, lateral drainage of water in drainage layers, leachate recirculation, vertical percolation of soil water, and leakage through hydraulic barriers (geomembranes, clay, or geomembrane/clay composite liners).

Engineering documentation of HELP is provided by Schroeder et al. (1994). We will not attempt to repeat the documentation here. Instead, we will provide an overview of HELP's capability and discuss the key technical components of the model. The HELP program contains a number of default values for soil and other parameters, which can prove to be very helpful even for hand analyses. Tables of critical default values are included herein.

4.5.1 Design Profile

A schematic view of the typical profile that HELP was designed to simulate is shown in Figure 4.7. The profile is divided into up to three subprofiles (cover, waste, and bottom liner system) to simulate a landfill. For purposes of this book, attention is focused on the cover.

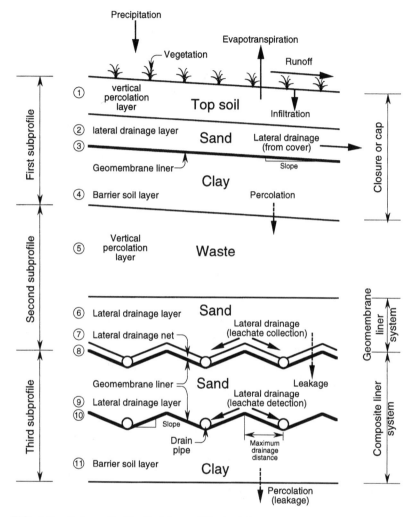

FIG. 4.7. Schematic Profile View of Typical Solid Waste Landfill

The layers that are analyzed with HELP are categorized by the hydraulic function that they perform. Four types of layers are available, as summarized in Table 4.8.

4.5.1.1 Vertical Percolation Layer.

A vertical percolation layer is any layer permitting vertical movement of water (downward, due to gravity or upward, due to evapotranspiration) within it, and not serving as a lateral drainage layer. Low-permeability soils are normally treated as a vertical

TABLE 4.8. Four Types of Layers Allowed in HELP Computer Code

Type of Layer	Hydraulic Characteristics
Vertical Percolation Layer	Flow in this layer is strictly vertical (downward due to gravity or upward due to evapotranspiration). Hydraulic conductivity at saturation is typically in the range of 10^{-5} to 10^{-8} m/s.
Lateral Drainage Layer	This layer promotes lateral drainage to collection systems, e.g., drains at the perimeter of the cover. Hydraulic conductivity is typically $>10^{-4}$ m/s and the underlying layer is normally a liner.
Barrier Soil Liner	Barrier soil liners are low-permeability soils, normally a compacted clay liner (CCL) or geosynthetic clay liner (GCL). The barrier soil layer normally has a hydraulic conductivity less than 10^{-8} to 10^{-9} m/s.
Geomembrane Liner	Geomembranes can be of any type. They are assumed to permit leakage via vapor diffusion, manufacturing flaws (pinholes), and installation defects (e.g., flaws).

percolation layer, if desired. Examples of layers that are normally treated as a vertical percolation layer are the top soil, protection layer, gas collection layer, foundation soil, and waste.

Water flow in a vertical percolation layer is assumed to occur in accord with Darcy's formula:

$$f = ki \qquad (4.35)$$

where "f" is the flux (volume of flow per unit cross-sectional area per unit time), "k" is the hydraulic conductivity (which varies with the water content of the soil), and "i" is the dimensionless hydraulic gradient (i.e., change in total hydraulic head per unit distance along the path of water flow). The total hydraulic head (H) is equal to the sum of pressure (H_p) and elevation (z) heads:

$$H = H_p + z \qquad (4.36)$$

Consider the case in which the water content is constant throughout the vertical percolation layer. The pressure head in an unsaturated soil is a function of water content. The drier the soil, the lower the water content and the more negative the pressure (or capillary) head. Thus, if the water content is uniform throughout a vertical percolation layer, the pressure head is uniform everywhere. As shown in Figure 4.8, the change in head between two points in a vertical percolation layer under these assumptions is 1.0.

Program HELP assumes that the pressure head is uniform throughout most vertical percolation layers and, therefore, that gravity drainage occurs at unit hydraulic gradient. In situations where head can build up on top of

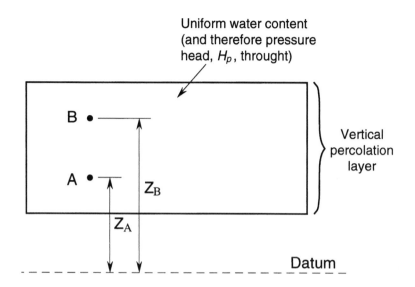

Hydraulic gradient (i) between A and B:

$$i = \frac{\Delta H}{L} = \frac{\left[Z_B + (H_p)_B\right] - \left[Z_A + (H_p)_A\right]}{Z_B - Z_A}$$

$$= \frac{Z_B - Z_A}{Z_B - Z_A} = 1$$

FIG. 4.8.   *Unit Hydraulic Gradient in Vertical Percolation Layer with Uniform Water Content*

the layer (e.g., a low-permeability vertical percolation layer or a soil barrier layer), the hydraulic gradient is computed as follows:

$$i = 1 + (H_w/L) \tag{4.37}$$

where H_w is the head of water on top of the layer and L is the thickness of the layer.

The hydraulic conductivity (k) of the unsaturated vertical percolation layer is computed from Campbell's (1974) equation:

$$k = k_{sat}[(\theta - \theta_r)/(\theta_{sat} - \theta_r)]^{3+2/\lambda} \tag{4.38}$$

where k_{sat} is the hydraulic conductivity of the soil at saturation, θ is the volumetric water content, θ_r is the residual water content (i.e., water content at very large suction), θ_{sat} is the volumetric water content at saturation (θ_{sat} is equal to the porosity), and λ is a dimensionless pore-size distribution index. The residual water content may be assumed to equal the water content at the wilting point, although HELP estimates the residual water content, which typically falls in the range of 0.01 to 0.03. The pore-size distribution index is determined from laboratory soil moisture retention data. In the laboratory, soils are wetted or dried to a range of water contents, and the corresponding soil water potential, ψ, is measured. (For those more familiar with the term, suction is defined as the negative of soil water retention.) A "soil moisture characteristic curve" is then prepared, which is simply a graph of θ versus ψ. The $\theta-y$ data are graphed in log–log, and λ is the slope of the log–log plot (Campbell 1974).

Soil moisture characteristic curves have been measured on thousands of soils, and λ has been found to vary with the texture of the soil. It is common practice to estimate, rather than measure, λ. Given the inevitable uncertainty in k_{sat}, which is often plus or minus an order of magnitude or more, one usually cannot justify the expense of measuring λ. And more often than not, the actual materials of construction are selected by the construction contractor and, therefore, are not available for testing during design.

The method of calculating the downward flux of water in the unsaturated vertical percolation layer is approximate. Far more vigorous analytic techniques are available that more carefully compute hydraulic gradients and consider vapor and thermal transport mechanisms. However, computer codes that account for unsaturated flow more rigorously tend to be difficult to use because of their complexity and, therefore, are rarely employed for water balance analyses. Nevertheless, HELP is not considered a particularly accurate simulation program for covers that are located in arid areas, where the subtleties of unsaturated moisture movement can dominate the water balance.

4.5.1.2 Lateral Drainage Layer.
Lateral drainage layers may consist of granular soils or geosynthetic materials. Vertical drainage in a lateral drainage layer is modeled in the same manner as a vertical percolation layer. However, lateral flow in the saturated zone at the base of the lateral drainage layer is allowed.

Unconfined lateral flow in the drainage layer is modeled using Darcy's formula, assuming continuity and employing the Depuit-Forcheimer assumptions (seepage parallel to the slope of the layer and hydraulic gradient proportional to the slope of the water table surface). The algorithm used by HELP is reasonably rigorous and accurate. The accuracy with which the hydraulic conductivity of the lateral drainage is defined and the frequency of precipitation data (daily rather than hourly) limit the overall accuracy of the calculations.

One of the important calculated parameters is the maximum head of the water in the drainage layer. A drainage layer is often included more to maintain stability of slopes then to reduce infiltration of water. If head builds up (e.g., rises above the top of the drainage layer or, worse yet, to the top of the cover itself), the water pressures could trigger slope movements. For this reason, it is important to examine the calculated maximum head in the drainage layer carefully. Also, because of the limitation of analysis with daily precipitation data, as discussed in Section 4.4.2, the HELP model may be unconservative in prediction of peak heads and flow rates. Hourly precipitation data are recommended to estimate peak flows, per Section 4.4.2.

As mentioned earlier, the key input parameter, in terms of lateral drainage layer, is the hydraulic conductivity of the drainage layer material. Designers are urged to use caution and to be conservative. The designer normally specifies the hydraulic conductivity of the drainage material, and this value is verified as part of the construction quality assurance process. However, the specified value should be greater than the minimum acceptable value because of the many changes that can occur in the drainage layer during or after construction, nearly all of which tend to drive the hydraulic conductivity downward. Some of the hydraulic-conductivity-reducing processes of particular concern include crushing or grinding of granular materials during construction (generating fines that reduce hydraulic conductivity), deposition of dust during construction, tracking of fines into the drainage material by the wheels or treads of construction vehicles, deposition of precipitates from the soil water, growth of micro-organisms, creep that reduces porosity, and clogging by fines from adjacent soils migrating into the drainage material. As a minimum, the authors recommend specifying a hydraulic conductivity (or transmissivity for geosynthetics) that is at least 5 to 10 times larger than the required value. The designer is also reminded that drainage layers always should meet filter criteria for adjacent materials, or a filter (soil or geosynthetic) should be provided.

4.5.1.3 Low-Permeability Soil Barrier Layer.
Compacted clay liners (CCLs) and geosynthetic clay liners (GCLs) are frequently used as hydraulic barrier layers in covers. Program HELP computes leakage through the soil using Darcy's formula and Equation 4.32 for hydraulic gradient. The soil is assumed to be saturated, i.e., to have no capacity to store water without drainage occurring. Leakage through the CCL or GCL is assumed to occur whenever there is a head of water on top of the barrier ($h_w > 0$ in Equation 4.32).

In many situations, particularly when the soil liner is located within 1 m of the surface of the cover and there is no geomembrane overlying the clay, the low-permeability soil layer will probably desiccate at times, invalidating the assumption of continuous saturation. To model this process, the low-permeability soil layer can be treated as a vertical percolation layer. Also, clay liners are almost never saturated with water at the time of construction, so the liners must first absorb some water before drainage is initiated.

4.5.1.4 Geomembrane Layer.
Geomembranes are widely and routinely used in well-engineered covers. Geomembranes can be extremely effective hydraulic barriers and can withstand many of the forces (e.g., differential settlement and freeze/thaw or wet/dry cycles) that are so destructive to clay liners.

Program HELP assumes that liquids can leak through geomembranes by three mechanisms: (1) vapor diffusion through the intact geomembrane; (2) leakage through manufacturing defects (pinholes); and (3) leakage through construction defects (mainly defects in seams). The equations are complex and involve a number of possible cases. The reader is referred to Schroeder et al. (1994) for details.

4.5.2 Default Properties

One of the useful aspects of the HELP model is that it contains default parameters for various soil and waste properties based upon data available for more than a thousand soils. Default properties are summarized in Table 4.9 for low-density soils (i.e., soils with low to moderate compaction) and in Table 4.10 for moderate- and high-density soils. In these two tables, the first column, labeled HELP, refers to the internal tracking (soil number) used by HELP for any particular soil. The second column indicates U.S. Department of Agriculture (USDA) classification, while the third column shows the Unified Soil Classification System (USCS) symbol (see Table 4.11 for an explanation of the symbols). The wilting point is defined as the volumetric water content at a suction of -15 bars, which is approximately the maximum suction that most plants can withstand before they wilt and die. Table 4.12 provides information on default waste characteristics, Table 4.13 presents information on saturated hydraulic conductivity of wastes, and Table 4.14 summarizes default material characteristics for various geosynthetic materials.

4.5.3 Method of Solution

Program HELP models both surface processes and subsurface processes. The surface processes include snowmelt, interception of rainfall by vegetation, surface runoff, and evaporation of water. The subsurface processes modeled are evaporation of water from the soil, transpiration of water by plants, vertical percolation of water through unsaturated soil, lateral drainage in drainage layers, and leakage of water through soil, geomembrane,

TABLE 4.9. Default Low Density Soil Characteristics

Soil Texture Class			Total Porosity (vol/vol)	Field Capacity (vol/vol)	Wilting Point (vol/vol)	Saturated Hydraulic Conductivity (cm/sec)
HELP	USDA	USCS				
1	CoS	SP	0.417	0.045	0.018	1.0×10^{-2}
2	S	SW	0.437	0.062	0.024	5.8×10^{-3}
3	FS	SW	0.457	0.083	0.033	3.1×10^{-3}
4	LS	SM	0.437	0.105	0.047	1.7×10^{-3}
5	LFS	SM	0.457	0.131	0.058	1.0×10^{-3}
6	SL	SM	0.453	0.190	0.085	7.2×10^{-4}
7	FSL	SM	0.473	0.222	0.104	5.2×10^{-4}
8	L	ML	0.463	0.232	0.116	3.7×10^{-4}
9	SiL	ML	0.501	0.284	0.135	1.9×10^{-4}
10	SCL	SC	0.398	0.244	0.136	1.2×10^{-4}
11	CL	CL	0.464	0.310	0.187	6.4×10^{-5}
12	SiCL	CL	0.471	0.342	0.210	4.2×10^{-5}
13	SC	SC	0.430	0.321	0.221	3.3×10^{-5}
14	SiC	CH	0.479	0.371	0.251	2.5×10^{-5}
15	C	CH	0.475	0.378	0.251	2.5×10^{-5}
21	G	GP	0.397	0.032	0.013	3.0×10^{-1}

or composite liners. Daily infiltration of water into the surface of the cover is determined indirectly from a surface water balance. Each day, infiltration is assumed to equal the sum of rainfall and snowmelt, minus the sum of runoff, surface storage (e.g., on the surfaces of plants), and surface evaporation (e.g., evaporation of water stored on the surfaces of plants).

The daily surface-water accounting procedure used in HELP is as follows. Snowfall and rainfall are added to the surface snow storage, if present, and then snowmelt plus excess storage of rainfall is computed. The total outflow from the snow cover is then treated as rainfall in the absence of a snow cover for the purpose of computing runoff. A rainfall-runoff relationship is used to calculate runoff. Surface evaporation is then computed, but surface evaporation is not allowed to exceed the sum of surface snow storage and intercepted rainfall. The snowmelt and rainfall that does not run off or evaporate is assumed to infiltrate the landfill. Computed infiltration in excess of the storage and drainage capacity of the soil is routed back to the surface and is added to the runoff or held as surface storage.

The subsurface processes modeled by HELP are as follows. The first subsurface process considered is evaporation of water from the soil. Next, transpiration of water from the evaporative zone of the by plants is computed. Other processes are modeled using a time step varying from 30

TABLE 4.10. Moderate and High Density Default Soil Characteristics

Soil Texture Class			Total Porosity (vol/vol)	Field Capacity (vol/vol)	Wilting Point (vol/vol)	Saturated Hydraulic Conductivity (cm/sec)
HELP	USDA	USCS				
22	L (Moderate)	ML	0.419	0.307	0.180	1.9×10^{-5}
23	SiL (Moderate)	ML	0.461	0.360	0.203	9.0×10^{-6}
24	SCL (Moderate)	SC	0.365	0.305	0.202	2.7×10^{-6}
25	CL (Moderate)	CL	0.437	0.373	0.266	3.6×10^{-6}
26	SiCL (Moderate)	CL	0.445	0.393	0.277	1.9×10^{-6}
27	SC (Moderate)	SC	0.400	0.366	0.288	7.8×10^{-7}
28	SiC (Moderate)	CH	0.452	0.411	0.311	1.2×10^{-6}
29	C (Moderate)	CH	0.451	0.419	0.332	6.8×10^{-7}
16	Liner Soil (High)		0.427	0.418	0.367	1.0×10^{-7}
17	Bentonite (High)		0.750	0.747	0.400	3.0×10^{-9}

minutes to 6 hours. For vertical percolation layers, a water balance is performed on each layer to determine the water content of the material. Hydraulic conductivity is computed from the water content, and then the amount of gravity drainage (if any) is determined. For lateral drainage layers, a water balance is used to determine whether the drainage layer is saturated at any point, and if so, lateral drainage is computed for that portion of the layer that is saturated. Vertical percolation is assumed to occur in the lateral drainage layer above the zone of saturation—the same equations employed for analyzing gravity drainage in vertical percolation layers are used to analyze vertical flow above the saturated zone in lateral drainage layers. Soil barrier layers are assumed to be continuously saturated and, therefore, no water balance is performed for them. Leakage is computed from the hydraulic properties of the drainage layer and the amount of head acting on the barrier layer. Leakage through geomembranes is computed from vapor diffusion, leakage through pinholes, and leakage through manufacturing defects.

TABLE 4.11.   Default Soil Texture Abbreviations

US Department of Agriculture	Definition
G	Gravel
S	Sand
Si	Silt
C	Clay
L	Loam (sand, silt, clay, and humus mixture)
Co	Coarse
F	Fine

Unified Soil Classification System	Definition
G	Gravel
S	Sand
M	Silt
C	Clay
P	Poorly Graded
W	Well Graded
H	High Plasticity or Compressibility
L	Low Plasticity or Compressibility

TABLE 4.12.   Default Waste Characteristics

Waste Identification		Total Porosity (vol/vol)	Field Capacity (vol/vol)	Wilting Point (vol/vol)	Saturated Hydraulic Conductivity (cm/sec)
HELP	Waste Material				
18	Municipal Waste	0.671	0.292	0.077	1.0×10^{-3}
19	Municipal Waste with Channeling	0.168	0.073	0.019	1.0×10^{-3}
30	High-Density Electric Plant Coal Fly Ash*	0.541	0.187	0.047	5.0×10^{-5}
31	High-Density Electric Plant Coal Bottom Ash*	0.578	0.076	0.025	4.1×10^{-3}
32	High-Density Municipal Solid Waste Incinerator Ash**	0.450	0.116	0.049	1.0×10^{-2}
33	High-Density Fine Copper Slag**	0.375	0.055	0.020	4.1×10^{-2}

*All values, except saturated hydraulic conductivity, are at maximum dry density. Saturated hydraulic conductivity was determined in-situ.
**All values are at maximum dry density. Saturated hydraulic conductivity was determined by laboratory methods.

TABLE 4.13. Saturated Hydraulic Conductivity of Wastes

Waste Material	Saturated Hydraulic Conductivity (cm/sec)*	Reference
Stabilized Incinerator Fly Ash	8.8×10^{-5}	Poran and Ahtchi-Ali (1989)
High-Density Pulverized Fly Ash	2.5×10^{-5}	Swain (1979)
Solidified Waste	4.0×10^{-2}	Rushbrook et al. (1989)
Electroplating Sludge	1.6×10^{-5}	Bartos and Palermo (1977)
Nickel/Cadmium Battery Sludge	3.5×10^{-5}	Bartos and Palermo (1977)
Inorganic Pigment Sludge	5.0×10^{-6}	Bartos and Palermo (1977)
Brine Sludge—Chlorine Production	8.2×10^{-5}	Bartos and Palermo (1977)
Calcium Fluoride Sludge	3.2×10^{-5}	Bartos and Palermo (1977)
High Ash Papermill Sludge	1.4×10^{-6}	Perry and Schultz (1977)

*Determined by laboratory methods.

Program HELP is widely used in engineering practice and is a routine part of the design practice. The program is convenient to use, particularly in terms of default soil and waste properties, and weather data contained within the program. However, users should be aware of the fact that much of the default weather data is for the period 1974–1977, which was an unusually dry period in certain parts of the United States (particularly the western United States). Because of its convenience and built-in parameters, the program is also easy to mis-use, in part because one does not have to

TABLE 4.14. Default Geosynthetic Material Characteristics

Geosythetic Material Description		Saturated Hydraulic Conductivity (cm/sec)
HELP	Geosynthetic	
20	Drainage Net (0.5 cm)	$1.0 \times 10^{+1}$
34	Drainage Net (0.6 cm)	$3.3 \times 10^{+1}$
35	High Density Polyethylene (HDPE) Geomembrane	2.0×10^{-13}
36	Low Density Polyethylene (LDPE) Geomembrane	4.0×10^{-13}
37	Polyvinyl Chloride (PVC) Geomembrane	2.0×10^{-11}
38	Butyl Rubber Geomembrane	1.0×10^{-12}
39	Chlorinated Polyethylene (CPE) Geomembrane	4.0×10^{-12}
40	Hypalon or Chlorosulfonated Polyethylene (CSPE) Geomembrane	3.0×10^{-12}
41	Ethylene-Propylene Diene Monomer (EPDM) Geomembrane	2.0×10^{-12}
42	Neoprene Geomembrane	3.0×10^{-12}

understand very much about the water balance process or hydraulic properties of the components of covers to run the program.

One of the most useful applications of HELP is for sensitivity analyses in which the values of various key parameters (such as the thickness and hydraulic conductivity of various layers within the cover) are varied to evaluate the impact upon performance of the cover. One way that the program can be misused is to demonstrate whether or not leachate will be generated during the period in which a landfill is uncovered—whether or not leachate is produced depends almost entirely on the assumption about the initial moisture content of the waste (whether or not it is close to field capacity), and this type of information is usually known with poor accuracy. Thus, in a sense, one could get just about any answer from HELP that one wants, depending on the key assumption about the initial water content of the waste.

For persons who review the calculations from HELP, it is very important to examine carefully the inputs to the model. Simply to run the program does not necessarily mean that appropriate conclusions are reached. Appropriate conclusions are reached only if appropriate inputs are made to the model. Reviewers should carefully examine the input precipitation to confirm that appropriate meteorological conditions have been modeled (e.g., an abnormally wet year if the purpose is to calculate maximum flow rates) and that worst-case assumptions have been made, when appropriate.

4.6 DESIGN PERCOLATION RATE

One of the problems with regulations concerning covers for landfills is that there are no standards for the allowable rate of water percolation through a cover cross section. For risk-based corrective action projects, the percolation rate is usually computed and then used to predict impacts on groundwater as a part of the process of evaluating health risks.

The authors believe that the landfill industry would be well served by implementing standards for expected percolation rates through cover systems. We have effectively done this in Chapter 3, suggesting different cover profiles for different percolation objectives. At the very least, the authors believe that designers should calculate the expected percolation rate through the cover, e.g., using program HELP. Armed with this calculation, the designer could (and should) consider alternative designs that would result in larger or smaller percolations, depending on the objectives established for each project.

It is naive to think that covers will yield zero percolation. Some percolation should be expected. As a starting point, one may wish to note that continuous leakage at unit hydraulic gradient through an intact clay liner with a hydraulic conductivity of 1×10^{-7} cm/s yields a percolation rate of about 25 mm/year (1 inch/year). Non-engineered covers composed of a thin layer of soil probably yield percolation rates of 100 to 300 mm/year in humid areas. Well engineered covers almost certainly will reduce percolation rates to no more than 1 to 10 mm/year, and probably far less (≪1 mm/year) for the most sophisticated designs employing a thick cover soil, a drainage layer, and a geomembrane/clay composite barrier. Only by analyzing the expected performance of covers can we make informed decisions about how sophisticated and complex the cover should be.

4.7 REFERENCES

Bartos, M. J., and Palermo, M. R. (1977). "Physical and Engineering Properties of Hazardous Industrial Wastes and Sludges," Technical Resource Document, EPA-600/2-77-139, U.S. Army Engineer Waterways Experiment Station, Vicksburg, MS.

Campbell, G. S. (1974). "A Simple Method for Determining Unsaturated Hydraulic Conductivity from Moisture Retention Data," Soil Science, 117(6):311–314.

Fenn, D. G., Hanley, K. J., and DeGeare, T. V. (1975). "Use of the Water Balance Method for Predicting Leachate Generation from Solid Waste Disposal Sites," U.S. Environmental Protection Agency, EPA/530/SW-168, Washington, D.C., 40 pgs.

Giroud, J. P., and Bonaparte, R. (1989). "Leakage through Liners Constructed with Geomembrane Liners—Parts I and II and Technical Note," Journal Geotextiles and Geomembranes, 8(1): pp. 27–67; 8(2): pp. 71–111; 8(4): pp. 337–340.

Kmet, P. (1982). "EPA's 1995 Water Balance Method—Its Use and Limitations," Wisconsin Department of Natural Resources, Madison, WI.

Perry, J. S., and Schultz, D. I. (1977). "Disposal and Alternate Uses of High Ash Paper-Mill Sludge," Proceedings of the 1977 National Conference on Treatment and Disposal of Industrial Wastewaters and Residues, University of Houston, Houston, TX.

Poran, C. J., and Actchi-Ali, F. (1989). "Properties of Solid Waste Incinerator Fly Ash," Journal of Geotechnical Engineering, 115(8):1118–1133.

Ritchie, J. T. (1972). "A Model for Predicting Evaporation from a Row Crop with Incomplete Cover," Water Resources Research, 8(5):1204–1213.

Rushbrook, P. E., Baldwin, G., and Dent, C. B. (1989). "A Quality-Assurance Procedure for Use at Treatment Plants to Predict the Long-Term Suitability of Cement-Based Solidified Hazardous Wastes Deposited in Landfill Sites," Environmental Aspects of Stabilization and Solidification of Hazardous and Radioactive Wastes, ASTM STP 1033, P. L. Cote and T. M. Gilliam, Eds., American Society for Testing and Materials, Philadelphia.

Schroeder, P. R., Dizier, T. S., Zappi, P. A., McEnroe, B. M., Sjostrom, J. W., and Peyton, R. L. (1994). "The Hydrologic Evaluation of Landfill Performance (HELP) Model: Engineering Documentation for Version 3," EPA/600/R-94/168b, U.S. Environmental Protection Agency, Risk Reduction Engineering Laboratory, Cincinnati, OH, 116 pgs.

Swain, A. (1979). "Field Studies of Pulverized Fuel Ash in Partially Submerged Conditions," *Proceedings of the Symposium of the Engineering Behavior of Industrial and Urban Fill*, The Midland Geotechnical Society, University of Birmingham, Birmingham, England, D49–D61.

Thiel, R. S., and Stewart, M. G. (1993). "Geosynthetic Landfill Cover Design Methodology and Construction Experience in the Pacific Northwest," Geosynthetics '93 Conference Proceedings, IFAI, St. Paul, MN, pp. 1131–1144.

Thornthwaite, C. W., and Mather, J. R. (1955). "The Water Balance," Drexel Institute of Technology, Publications in Climatology, Vol. 8, No. 1, Philadelphia, PA.

CHAPTER 5

SLOPE STABILITY OF FINAL COVER SYSTEMS

With increasing needs to maximize landfill air space (by virtue of economics, logistics, politics, etc.), the slopes of final covers tend toward being relatively steep, rather than being relatively flat. Slopes of 3(H)-to-1(V), i.e., 18.4 deg. with the horizontal, are somewhat common and 2(H)-to-1(V), i.e., 26.6 deg. with the horizontal, have also been utilized. Couple these steep slopes with the realization that low interface shear strength inclusions (such as geomembranes, hydrated GCLs, and/or wet-of-optimum CCLs) are oriented precisely in the direction of a potential slide, and the necessity for careful slope stability analyses should be obvious. Hence, the reason for this chapter.

5.1 OVERVIEW

With geomembranes, hydrated GCLs and/or wet-of-optimum CCLs used as barrier layer components in a final cover situation, potential shear planes exist at a number of interfaces. Even further, geosynthetic drainage systems (for water drainage above and gas transmission below) can create additional potential shear planes. A number of slides parallel to the slope angle have occurred and have been reported in the open literature. The most common situations appear to be the following:

- Cover soil sliding off the upper surface of a smooth geomembrane.
- Cover soil with an underlying geotextile or drainage geocomposite sliding off the upper surface of a smooth geomembrane.
- Cover soil, drainage materials and underlying geomembrane sliding off the upper surface of the underlying soil, e.g., a wet-of-optimum CCL.
- Cover soil, drainage materials and underlying geomembrane sliding off the upper surface of a underlying hydrated GCL, par-

ticularly if the upper surface of the GCL is a woven slit film geotextile.

From the perspective of a slope stability analysis, the actual or potential shear plane is generally linear, parallel to the slope angle and along the surface having the lowest interface shear strength. The analysis is straightforward and within the state-of-the-practice. The analyses used herein are based upon limit equilibrium principles, however, it should be recognized that finite element methods have also been used for the same class of problems (see Wilson-Fahmy and Koerner 1993). Limit equilibrium analysis is a methodology which requires material-specific shear strength properties that are obtained from laboratory tests simulating the site-specific situation as closely as possible. In this regard, the results of direct shear tests will be seen to be the most significant input in the analysis. The accuracy of the slope stability analysis is limited by the accuracy of the measured interface shear strength more than any other single factor.

The result of slope stability analyses, of the type to be described in this chapter, is a global factor of safety (FS). This FS value must be viewed in light of the significance of a potential failure. For example, a temporary landfill cover might be designed for a factor of safety of 1.2 to 1.3, while a final cover under the same circumstances would generally require a value of 1.4 to 1.5. This, of course, depends upon site-specific conditions and (usually) a review by the appropriate regulatory agency.

In the sections to follow in this chapter, the general principles of limit equilibrium analysis will be presented. Details and nuances of direct shear testing will be included. Various slope stability scenarios will follow, with details on both constant thickness cover soils and live load considerations. To offset potentially low FS values, procedures using tapered thickness cover soils and/or using geosynthetic reinforcement will be presented. Seepage considerations will form a separate section, as will seismic considerations. The summary will give design suggestions on approaches to minimize slope stability concerns of final covers, and some alternative strategies. It will also present the authors' recommendations for minimum FS values under different waste-specific and risk-specific situations.

5.2 GEOTECHNICAL PRINCIPLES AND ISSUES

As mentioned previously, the potential failure surface for final covers is usually linear with an overlying cover soil sliding with respect to the lowest interface friction layer in the underlying cross section. The potential failure plane's linear character allows for a straightforward calculation without the need for trial center locations and different radii, as with soil slopes analyzed by rotational failure surfaces. Furthermore, full static equilibrium can be achieved without solving simultaneous equations.

5.2.1 Limit Equilibrium Concepts

The free body diagram of an *infinitely long slope* with uniformly thick cohesionless cover soil on an incipient planar shear surface, like the upper surface of a geomembrane, is shown in Figure 5.1. The situation can be treated quite simply.

By taking a force summation parallel to the slope and comparing the resisting force to the driving or mobilizing force, a global factor of safety (FS) results for a cohesionless (i.e., purely frictional) interface as follows:

$$FS = \frac{\sum \text{Resisting Forces}}{\sum \text{Driving Forces}}$$

$$= \frac{N \tan \delta}{W \sin \beta} = \frac{W \cos \beta \tan \delta}{W \sin \beta}$$

Hence

$$FS = \frac{\tan \delta}{\tan \beta} \qquad (5.1)$$

Thus it is seen that the FS value is the ratio of the tangent of the interface friction angle of the cover soil with the upper surface of the geomembrane (δ) to the tangent of the slope angle of the soil beneath the geomembrane (β). As simple as this analysis is, its teachings are very significant, for example:

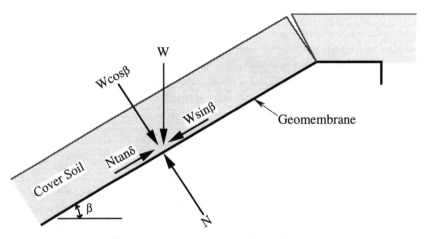

FIG. 5.1.   *Limit Equilibrium Forces Involved in Finite Slope Analysis for Uniformly Thick Cohesionless Soil*

- To obtain an accurate FS value, an accurately determined laboratory δ value is absolutely critical. The accuracy of the analysis is only as good as the accuracy of the laboratory-obtained δ value.
- For low δ values, the resulting soil slope angle will be proportionately low. For example, for a δ value of 20 deg., and a required FS value of 1.5, the maximum slope angle is 14 deg. This is equivalent to a 4(H)-on-1(V) slope, which is relatively flat. Furthermore, many geomembranes have even lower δ values, e.g., 10 to 15 deg.
- This simple formula has driven geosynthetic manufacturers to develop products with high δ values, i.e., textured geomembranes, thermally bonded drainage geocomposites, internally reinforced GCLs, etc.

The interface shear strength is defined as the shear strength at failure on a particular surface (interface). With many, but not all, interfaces, the interface shear strength may be purely frictional, but in others, there may be "cohesion" or "adhesion." Regardless of the presence of adhesion, the interface friction angle δ often varies with normal stress. It is critical to define δ at the appropriate normal stress expected in the field. If tan δ is determined as the shear strength at normal stress σ_n (where σ_n is the appropriate normal stress for the field situation) divided by σ_n then δ is said to be a "secant friction angle." If defined in this way, any adhesion can be ignored in calculating the factor of safety, because tan δ multiplied by σ_n will yield the correct shear strength, inclusive of the effects of adhesion.

The issue of peak versus residual δ is addressed later. Unfortunately, the above analysis is too simplistic to use in most practical situations. For example, the following situations cannot be accommodated:

- a finite-length slope with the incorporation of a passive soil wedge in the analysis,
- the incorporation of equipment loads on the slope,
- the use of tapered cover soils thickness,
- veneer reinforcement of the cover soil using geogrids or high-strength geotextiles,
- consideration of seepage forces in the cover soil, or
- consideration of seismic forces acting on the cover soil.

These situations will be treated in subsequent sections. For each situation, the essence of the theory will be presented, followed by the necessary design equations. This will be followed, in each case, with a design chart and an example problem. First, however, the issue of interface shear testing is discussed.

5.2.2 Interface Shear Testing

The interface shear strength of a cover soil with respect to the underlying material (often a geomembrane) is critical to properly analyzing the stability of the cover soil. This value of interface shear strength is obtained by laboratory testing of the project-specific materials at the site-specific conditions. By project-specific materials, we mean sampling of the candidate geosynthetics to be used at the site, as well as the cover soil (or other soil, such as CCL) at its targeted density and moisture content. By site-specific conditions, we mean testing at the anticipated normal stresses, moisture conditions, temperature extremes (high and/or low), strain rates and total deformation values. *Note, that it is completely inappropriate to use literature values of interface shear strengths for final cover design.* Literature values may be used for preliminary assessments, but the values must be confirmed during design or during CQA.

While the above list of items is formidable, at least the type of test is established. It is the direct shear test which has been used in geotechnical engineering testing for many years. The test has been adapted to evaluate geosynthetics in the United States and Germany as ASTM D5321 and DIN 60500 respectively.

In conducting a direct shear test on a specific interface, one typically performs three replicate tests with the only variable being three different values of normal stress. The middle value is usually targeted to the site-specific condition, with lower and higher values of normal stress covering the range of possible values. These three tests result in a set of shear displacement versus shear stress curves, see Figure 5.2a. From each curve, a peak shear strength (τ_p) and a residual shear strength (τ_r) is obtained. As a next step, these shear strength values, together with their respective normal stress values, are plotted on a Mohr-Coulomb stress space, as in Figure 5.2b. The points are then connected (usually with a straight line), and the two fundamental shear strength parameters are obtained. These shear strength parameters are:

δ = the angle of shearing resistance, peak and/or residual, of the two opposing surfaces (generally called the interface friction angle)

c_a = the adhesion of the two opposing surfaces, peak and/or residual (synonymous with cohesion when testing fine grained soils against one another)

These two parameters constitute the equation of a straight line, which is the Mohr-Coulomb failure criterion common to geotechnical engineering.

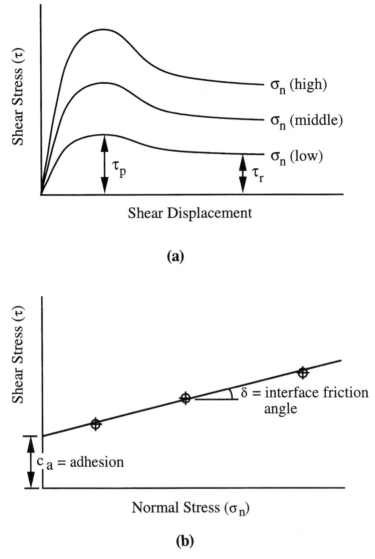

(a)

(b)

FIG. 5.2.   Direct Shear Test Results and Method of Analysis to Obtain Shear Strength Parameters

The concept is readily adaptable to geosynthetic materials in the following form:

$$\tau_p = c_{ap} + \sigma_n \tan \delta_p \qquad (5.2a)$$

$$\tau_r = c_{ar} + \sigma_n \tan \delta_r \qquad (5.2b)$$

The upper limit of "δ_p" when soil is involved as one of the interfaces, is "ϕ," the angle of shearing resistance of the soil component. The upper limit of the "c_{ap}" value is "c," the cohesion of the soil component. In the slope stability analyses to follow, the "c_{ap}" or "c_{ar}" term, if one is present, will not be utilized. There must be a clear physical justification for use of such values when geosynthetics are involved. Only unique situations, such as textured geomembranes with physical interlocking or the bentonite component of a GCL, are valid reasons for including such a term.

To be noted is that residual strengths are equal, or lower, than peak strengths. The amount of difference is very dependent on the material, and no general guidelines can be given. Clearly, project-specific and material-specific direct shear tests must be performed to determine the appropriate values. Further, each direct shear test must be conducted to an adequate displacement to determine the residual behavior (see Stark and Poeppel 1994). The decision as to the use of peak or residual strengths is a very difficult one because of lack of information on typical interfacial displacements in the field. It is clearly a site-specific and materials-specific issue which is up to the designer and/or regulator. Even further, the use of peak values at the crest of the slope and residual values at the toe may be justified. As such, the analyses to follow will use the δ value with no subscript, thereby concentrating on the computational procedures rather than this particular detail. However, its importance should not be minimized.

Due to the physical structure of many geosynthetics, the size of the recommended shear box is quite large. It should be at least 300 mm by 300 mm, unless it can be shown that data generated by a smaller device contains no scale or edge effects, i.e., that no bias exists with a smaller shear box. The implications of such a large shear box should not be taken lightly. Some issues which should receive particular attention are the following:

- Unless it can be justified otherwise, the interface will usually be tested in a saturated state. Thus complete and uniform saturation over the entire area must be achieved. This is particularly necessary for GCLs (Daniel et al. 1993). Hydration takes relatively long in comparison to soils in conventional (smaller) testing shear boxes.

- Consolidation of soils (including CCLs and GCLs) in large shear boxes is similarly affected.
- Uniformity of normal stress over the entire area must be maintained during consolidation and shearing so as to avoid the occurrence of stress concentrations.
- The application of relatively low normal stresses, e.g., 10, 20 and 30 kPa, simulating typical cover soil thicknesses, challenges the accuracy of some commercially available shear box setups and monitoring systems, particularly the accuracy of pressure gages.
- Shear rates necessary to attain drained conditions (if that is the desired situation) are extremely slow, requiring long test times.
- Deformations necessary to attain residual strengths require large relative movement of the two respective halves of the shear box. So as not to travel over the edges of the opposing shear box sections, many devices have the lower shear box larger than 300 mm. However, with a lower shear box larger than the upper traveling section, new surface is constantly being added to the shearing plane. The influence of this factor is not clear in the response or in the subsequent behavior.
- The attainment of a true residual strength is difficult to achieve. ASTM D5321 states that one should "run the test until the applied shear force remains constant with increasing displacement." Many commercially available shear boxes have insufficient travel to reach this condition.
- The ring torsion shearing apparatus is an alternative device to determine true residual strength values, but it is not without its own problems. See Stark and Poeppel (1994) for information and data using this alternative test method.

5.2.3 Various Situations Encountered

There is a large variety of slope stability problems that may be encountered in analyzing and/or designing final covers of engineered landfills, abandoned dumps, and remediation sites. Perhaps the most common is a uniformly thick cover soil on a geomembrane covering the slope at a given and constant slope angle. This "standard" problem will be analyzed in the next section. A variation of this problem will include equipment loads used during placement of cover soil on the geomembrane. This variation will be solved with equipment moving up the slope and then moving down the slope.

When low FS values arise in the above problems, the designer has a number of options. Other than a geometric redesign of the slope, there are two options commonly used. These are to use a tapered cover soil thickness and/or to use geosynthetic reinforcement. This latter option is called "veneer reinforcement" in the literature and comes about by the inclusion of a geogrid or high strength geotextile within the cover soil. Both of these situations will be illustrated.

Unfortunately, cover soil failures have occurred, and perhaps the majority of the failures have been associated with seepage forces. Indeed, drainage above a geomembrane (or other barrier material) in the cover soil cross section must be accommodated to avoid the possibility of seepage forces. A section will be devoted to this class of slope stability problems.

Lastly, the possibility of seismic forces exists in earthquake-prone locations. If an earthquake occurs in the vicinity of an engineered landfill, abandoned dump, or remediation site, the seismic wave travels through the solid waste mass, reaching the upper surface of the cover. It then decouples from the cover soil materials, producing a horizontal force which must be appropriately analyzed. A section will be devoted to the seismic aspects of cover soil slope analysis as well.

5.3 COVER SOIL SLOPE STABILITY PROBLEMS

This section presents the analytic formulations, design curves, and example problems for a number of common slope stability situations. The standard problem of a uniformly thick cover soil is developed without, then with, equipment loading. When the resulting FS value is too low, the designer can select a number of options. For example, lowering of the slope angle, reduction of the slope length with intermediate berms, or the use of higher shear strength materials are possible design options. The analysis procedure is the same regardless of these decisions. Quite different strategies are the use of tapered cover soil thickness and/or the use of high strength geosynthetic inclusions. These two strategies will be developed later in this section, being fundamentally different design alternatives.

5.3.1 Slopes with Uniformly Thick Cover Soils

Figure 5.3 illustrates a uniformly thick cover soil on a geomembrane at a slope angle "β" which is of finite length. It includes a passive wedge at the base and a tension crack at the top. The analysis that follows is after Koerner and Hwu (1991), but comparable analyses are available from Giroud and Beech (1989) and McKelvey and Deutsch (1991). The symbols used in Figure 5.3 are defined below.

W_A = total weight of the active wedge
W_P = total weight of the passive wedge
N_A = effective force normal to the failure plane of the active wedge
N_P = effective force normal to the failure plane of the passive wedge
γ = unit weight of the cover soil

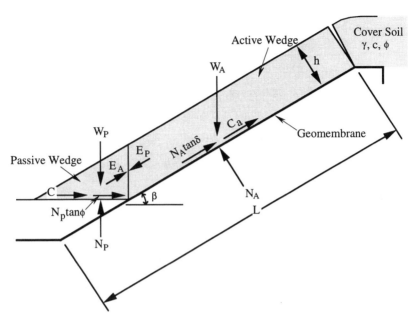

FIG. 5.3.   Limit Equilibrium Forces Involved in Finite Length Slope Analysis for Uniformly Thick Cover Soil

h = thickness of the cover soil

L = length of slope measured along the geomembrane

β = soil slope angle beneath the geomembrane

ϕ = friction angle of the cover soil

δ = interface friction angle between cover soil and geomembrane

C_a = adhesive force between cover soil of the active wedge and the geomembrane

c_a = adhesion between cover soil of the active wedge and the geomembrane

C = cohesive force along the failure plane of the passive wedge

c = cohesion of the cover soil

E_A = interwedge force acting on the active wedge from the passive wedge

E_P = interwedge force acting on the passive wedge from the active wedge

FS = factor of safety against cover soil sliding on the geomembrane

The expression for determining the factor of safety can be derived as follows:

Considering the active wedge,

$$W_A = \gamma h^2 \left(\frac{L}{h} - \frac{1}{\sin \beta} - \frac{\tan \beta}{2} \right) \qquad (5.3)$$

$$N_A = W_A \cos \beta \qquad (5.4)$$

$$C_a = c_a \left(L - \frac{h}{\sin \beta} \right) \qquad (5.5)$$

By balancing the forces in the vertical direction, the following formulation results:

$$E_A \sin \beta = W_A - N_A \cos \beta - \frac{N_A \tan \delta + C_a}{FS} \sin \beta \qquad (5.6)$$

Hence, the interwedge force acting on the active wedge is:

$$E_A = \frac{(FS)(W_A - N_A \cos \beta) - (N_A \tan \delta + C_a)\sin \beta}{\sin \beta(FS)} \qquad (5.7)$$

The passive wedge can be considered in a similar manner:

$$W_P = \frac{\gamma h^2}{\sin 2\beta} \qquad (5.8)$$

$$N_p = W_P + E_P \sin \beta \qquad (5.9)$$

$$C = \frac{(c)(h)}{\sin \beta} \qquad (5.10)$$

By balancing the forces in the horizontal direction, the following formulation results:

$$E_P \cos \beta = \frac{C + N_P \tan \phi}{FS} \qquad (5.11)$$

Hence, the interwedge force acting on the passive wedge is:

$$E_P = \frac{C + W_P \tan \phi}{\cos \beta(FS) - \sin \beta \tan \phi} \qquad (5.12)$$

By setting $E_A = E_P$, the following equation can be arranged in the form of $ax^2 + bx + c = 0$, which in our case use FS values, thus:

$$a(FS)^2 + b(FS) + c = 0 \qquad (5.13)$$

where

$$a = (W_A - N_A \cos \beta)\cos \beta$$

$$b = -[(W_A - N_A \cos \beta)\sin \beta \tan \phi + (N_A \tan \delta + C_a)\sin \beta \cos \beta$$
$$+ \sin \beta(C + W_P \tan \phi)]$$

$$c = (N_A \tan \delta + C_a)\sin^2\beta \tan \phi$$

$$(5.14)$$

The resulting FS value is then obtained from the following equation:

$$FS = \frac{-b + \sqrt{b^2 - 4ac}}{2a} \qquad (5.15)$$

When the calculated FS value falls below 1.0, a stability failure, with the cover soil sliding on the geomembrane, is to be anticipated. Thus, a value of greater than 1.0 must be sought as being the minimum factor of safety. How much greater than 1.0 the FS value should be is a design and/or regulatory issue. The issue of minimum allowable FS values, under different conditions, will be revisited at the end of this chapter.

In order to better illustrate the implications of Equations 5.13, 5.14, and 5.15, typical design curves for various FS values, as a function of slope angle and interface friction angle, are given in Figure 5.4. Note that the curves are developed specifically for the variables stated in the legend of the figure. Example problem #1 illustrates the use of the curves.

Example 1: Given: a 30 m long slope with a uniformly thick cover soil of 300 mm at a unit weight of 18 kN/m^3. The soil has a friction angle of 30 deg. and zero cohesion, i.e., it is a sand. The cover soil is on a geomembrane, as shown in Figure 5.3. Direct shear testing has resulted in an interface friction angle between the cover soil and geomembrane of 22 deg. with zero adhesion. What is the FS value at a slope angle of 3(H)-to-1(V), i.e., 18.4 deg?

Solution: Using the design curves of Figure 5.4 (which were developed for the exact conditions of the example problem), the resulting FS = 1.25.

Comment: In general, this is too low of a value for a final cover slope factor of safety, and a redesign is necessary. While there are many possible options for changing the geometry of the situation, the example will be revisited later in this section using tapered cover soil thickness and veneer reinforcement. Furthermore, this general problem will be used later, for comparison purposes to other cover soil slope stability situations.

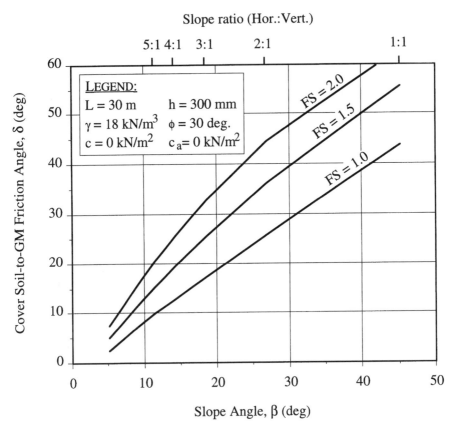

FIG. 5.4. *Design Curves for Stability of Uniform Thickness Cover Soils on Linear Failure Planes for Various Global Factors of Safety*

5.3.2 Incorporation of Equipment Loads

The placement of cover soil on a slope with a relatively low shear strength inclusion (like a geomembrane) should *always* be from the toe upward to the crest. Figure 5.5a shows the recommended method. In so doing, the gravitational forces of the cover soil and live load of the construction equipment are compacting previously placed soil and working with an ever-present passive wedge and stable lower portion beneath the active wedge. While it is prudent to specify low ground pressure equipment to place the soil, the reduction of the FS value from no equipment load while working up the slope will be seen to be nominal.

For soil placement down the slope, however, a stability analysis must add an additional dynamic stress into the solution. This stress decreases the

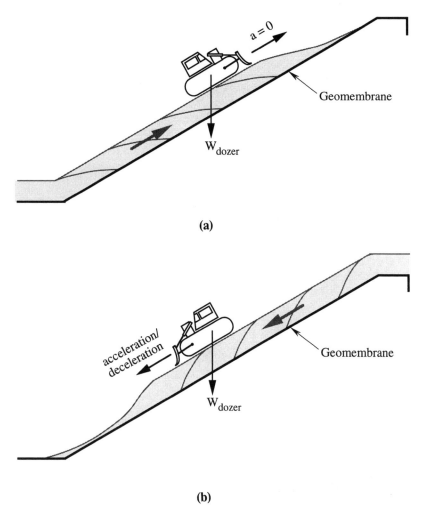

(a)

(b)

FIG. 5.5. *Construction Equipment Placing Cover Soil on Slopes Containing Geosynthetics*

FS value, and in some cases to a great extent. Figure 5.5b shows this procedure. Unless absolutely necessary, it is not recommended to place the cover soil in this manner. If it is necessary, the design must consider the dynamic force of the construction placement equipment.

For the **first case** of a bulldozer pushing cover soil up from the toe of the slope to the crest, the calculation uses the free body diagram of Figure 5.6a. The analysis uses a specific piece of construction equipment (like a

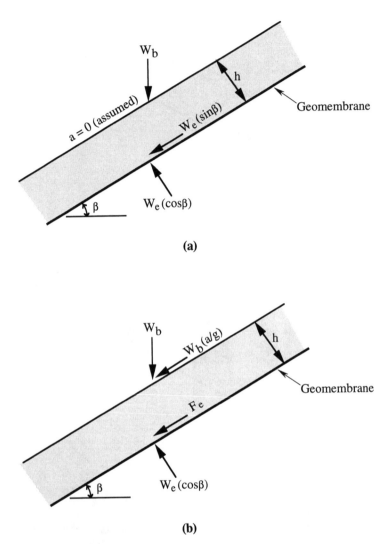

FIG. 5.6. *Additional Limit Equilibrium Forces due to Construction Equipment Moving on Cover Soil (See Fig. 5.3 for Gravitational Soil Forces that Remain the Same)*

bulldozer, characterized by its weight or ground-contact pressure), and dissipates this force or stress through the cover soil thickness to the surface of the geomembrane. A Boussinesq analysis is used (see Poulos and Davis 1974). This results in an equipment force per unit width as follows:

$$W_e = qwl \qquad (5.16)$$

where

W_e = equivalent equipment force per unit width at the geomembrane interface

q = $W_b/(2wb)$

W_b = actual weight of equipment (e.g., a bulldozer)

w = length of equipment track

b = width of equipment track

I = influence factor at the geomembrane interface (see Figure 5.7)

Upon determining the additional equipment force at the cover soil-to-geomembrane interface, the analysis proceeds as described in Section 5.3.1 for gravitational forces only. In essence, the equipment moving up the slope adds an additional term, W_e, to the W_A-force in Equation (5.3). Note, however, that this involves the generation of a resisting force as well. Thus, the net effect of increasing the driving force as well as the resisting force is somewhat neutralized, insofar as the resulting FS value is concerned.

Using this concept (the same equations used in Section 5.3.1 are used), typical design curves for various FS values, as a function of equivalent ground content pressures and cover soil thicknesses, are given in Figure 5.8. Note that the curves are developed specifically for the variables stated in the legend. Example problem #2(a) illustrates the use of the curves.

Example 2(a): Given: 30 m long slope with uniform cover soil of 300 mm thickness at a unit weight of 18 kN/m^3. The soil has a friction angle of 30 deg. and zero cohesion, i.e., it is a sand. It is placed on the slope using a bulldozer moving from the toe of the slope up to the crest. The bulldozer has a ground pressure of 30 kN/m^2 and tracks that are 3.0 m long and 0.6 m wide. The cover soil to geomembrane friction angle is 22 deg. with zero adhesion. What is the FS value at a slope angle of 3(H)-to-1(V), i.e., 18.4 deg?

Solution: This problem follows example #1 exactly, except for the addition of the bulldozer moving up the slope. Using the design curves of Figure 5.8 (which were developed for the exact conditions of the example problem), the resulting FS = 1.24.

Comment: While the resulting FS value is low, the result is best assessed by comparing it to example #1, i.e., the same problem except without the bulldozer. It is seen that the FS value has only decreased from 1.25 to 1.24. Thus, in general, a low ground contact pressure bulldozer placing cover soil up the slope does not significantly decrease the factor of safety.

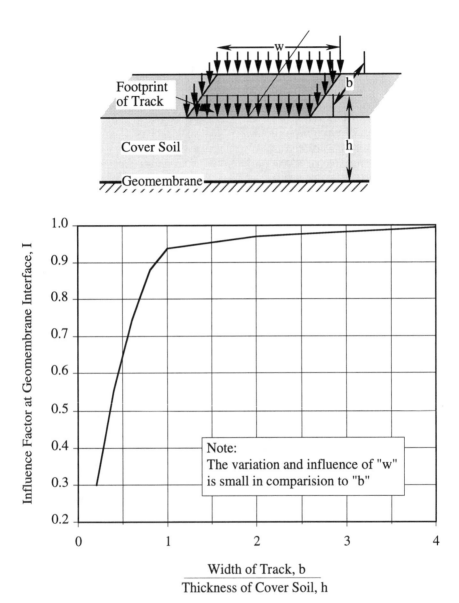

FIG. 5.7.   Value of Influence Factor, "I," for Use in Eq. (5.16) to Dissipate Surface Forces through Cover Soil to Geomembrane Interface [after Poulos and Davis (1974)]

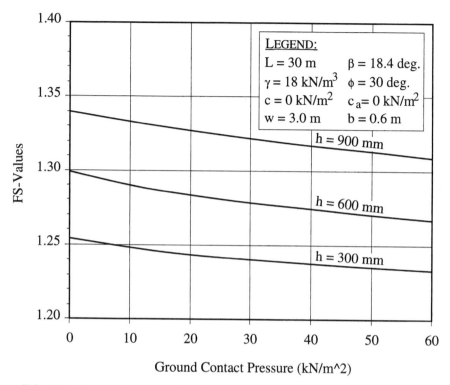

FIG. 5.8. *Design Curves for Stability of Different Thickness Cover Soils for Various Construction Equipment Ground Contact Pressure*

For the **second case** of a bulldozer pushing cover soil down from the crest of the slope to the toe, the analysis uses the force diagram of Figure 5.6b. While the weight of the equipment is treated as just described, an additional force due to acceleration (or deceleration) of the equipment must be added to the analysis. This analysis again uses a specific piece of construction equipment operated in a specific manner. It produces a force parallel to the slope equivalent to $W_b(a/g)$, where "W_b" is the weight of the bulldozer, a = acceleration of the bulldozer and g = acceleration due to gravity. Its magnitude is equipment operator-dependent and related to both the equipment speed and time to reach such a speed (see Figure 5.9).

The acceleration of the bulldozer, coupled with an influence factor "I" (from Figure 5.7), results in the dynamic force per unit width at the cover soil to geomembrane interface, "F_e." The relationship is as follows:

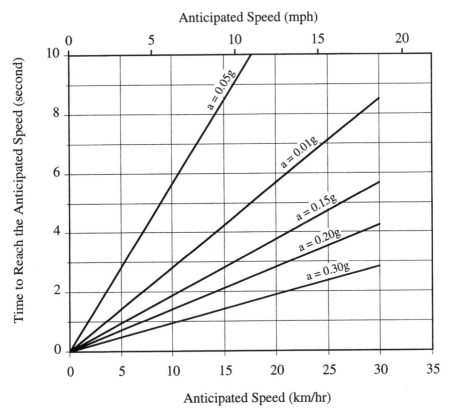

FIG. 5.9.　Graphic Relationship of Construction Equipment Speed and Rise Time to Obtain Acceleration

$$F_e = W_e \left(\frac{a}{g}\right) \tag{5.17}$$

where

F_e = dynamic force per unit width parallel to the slope at the geomembrane interface

W_e = equivalent equipment (bulldozer) force per unit width at geomembrane interface [recall Equation (5.16)]

β = soil slope angle beneath geomembrane

a = acceleration of the bulldozer

g = acceleration due to gravity

Using these concepts, the new force, parallel to the cover soil surface, is dissipated through the thickness of the cover sand to the interface of the geomembrane. Again, a Boussinesq analysis is used (see Poulos and Davis 1974). The expression for finding the FS value can now be derived as follows:

Considering the active wedge, and balancing the forces in the direction parallel to the slope, the following formulation results:

$$E_A + \frac{(N_e + N_A)\tan \delta + C_a}{FS} = (W_A + W_e)\sin \beta + F_e \qquad (5.18)$$

where

N_e = effective equipment force normal to the failure plane of the active wedge

$$= (W_e \cos \beta) \qquad (5.19)$$

Note that all the other symbols have been previously defined.

The interwedge force acting on the active wedge can now be expressed as:

$$E_A = \frac{(FS)[(W_A + W_e)\sin \beta + F_e] - [(N_e + N_A)\tan \delta + C_a]}{FS} \qquad (5.20)$$

The passive wedge can be treated in a similar manner, and the following formulation of the interwedge force acting on the passive wedge results:

$$E_P = \frac{C + W_P \tan \phi}{\cos \beta(FS) - \sin \beta \tan \phi} \qquad (5.21)$$

By setting $E_A = E_P$, the following equation can be arranged in the form of Equation (5.13), in which the "a," "b," and "c" terms are defined as follows:

$a = [(W_A + W_e)\sin \beta + F_e]\cos \beta$

$b = -\{[(N_e + N_A)\tan \delta + C_a]\cos \beta + [(W_A + W_e)\sin \beta + F_e]\sin \beta \tan \phi$

$\qquad + (C + W_P \tan \phi)\}$

$c = [(N_e + N_A)\tan \delta + C_a]\sin \beta \tan \phi \qquad (5.22)$

Finally, the resulting FS value can be obtained using Equation (5.15).

Using these concepts, typical design curves for various FS values, as a function of equipment ground contact pressure and equipment acceleration, can be developed (see Figure 5.10). Note that the curves are developed specifically for the variables stated in the legend. Example problem #2(b) illustrates the use of the curves.

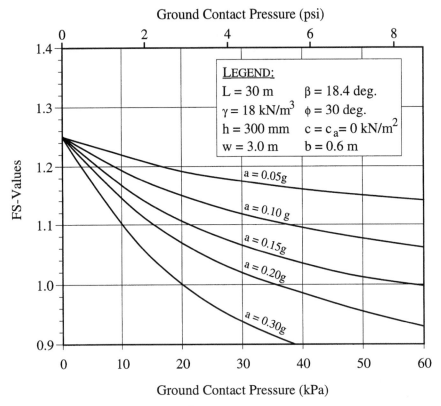

FIG. 5.10. Design Curves for Stability of Different Construction Equipment Ground Contact Pressures for Various Equipment Accelerations

Example 2(b): Given: a 30 m long slope with uniform cover soil of 300 mm thickness at a unit weight of 18 kN/m³. The soil has a friction angle of 30 deg. and zero cohesion, i.e., it is a sand. It is placed on the slope using a bulldozer moving from the crest down to the toe. The bulldozer has a ground contact pressure of 30 kN/m² and tracks that are 3.0 m long and 0.6 m wide. The estimated equipment speed is 20 km/hr and the time to reach this speed is 3.0 sec. The cover soil-to-geomembrane friction angle is 22 deg. with zero adhesion. What is the FS value at a slope angle of 3(H)-to-1(V), i.e., 18.4 deg? **Solution:** Using the design curves of Figures 5.9 and 5.10 (which were developed for the exact conditions of the example problem), we obtain the following:

- From Figure 5.9, with a velocity of 20 km/hr over 3.0 sec., the bulldozer's acceleration is 0.19g.
- From Figure 5.10 (which is developed for a cover soil thickness of 300 mm), with a value of a = 0.19g and q = 30 kPa, the resulting FS = 1.03.

Comment: This problem solution can now be compared to the previous two examples:

Ex. 1:	cover soil alone with no bulldozer	FS = 1.25
Ex. 2(a):	cover soil plus bulldozer moving up slope	FS = 1.24
Ex. 2(b):	cover soil plus bulldozer moving down slope	FS = 1.03

The inherent danger of a bulldozer moving down the slope is readily apparent. Note that the same result comes about by the bulldozer decelerating instead of accelerating. The sharp breaking action of the bulldozer is arguably the more severe condition due to the extremely short times involved when stopping forward motion. Clearly, only in unavoidable situations should the cover soil placement equipment be allowed to work down the slope. If it is unavoidable, an analysis should be made of the specific stability situation and the construction specifications should reflect the exact conditions made in the analysis. At the minimum, the ground contact pressure of the equipment should be stated along with suggested operator control of the cover soil placement operations.

5.3.3 Slopes with Tapered Thickness Cover Soils

One method available to the designer to increase the FS value of a slope is to taper the cover soil thickness from thick at the toe, to thin at the crest (see Figure 5.11). The FS value will increase in approximate proportion to the thickness of soil at the toe. The analysis for tapered cover soils includes the design assumptions of a tension crack at the top of the slope, the upper surface of the cover soil tapered at a constant angle "ω," and the earth pressure forces on the respective wedges oriented at the average of the surface and slope angles, i.e., the E forces are at an angle of $(\omega + \beta)/2$. The procedure follows that of the uniform cover soil thickness analysis. Again, the resulting equation is not an explicit solution for the FS, and must be solved indirectly. All symbols used in Figure 5.11 were previously defined (see Section 5.3.1) except the following:

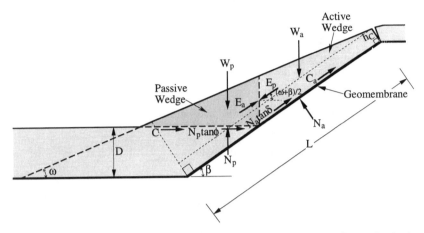

FIG. 5.11.   Limit Equilibrium Forces Involved in Finite Length Slope Analysis
for Tapered Thickness Cover Soil

D = thickness of the cover soil at bottom of the landfill
h_c = thickness of the cover soil at crest of the slope
y = $[L - (D/\sin \beta) - h_c \tan \beta](\sin \beta - \cos \beta \tan \omega)$
ω = finished cover soil slope angle

The expression for finding the factor of safety can be derived as follows:

Considering the active wedge,

$$W_A = \gamma \left[\left(L - \frac{D}{\sin \beta} - h_c \tan \beta \right) \left(\frac{y \cos \beta}{2} + h_c \right) + \frac{h_c^2 \tan \beta}{2} \right] \quad (5.23)$$

$$N_A = W_A \cos \beta \quad (5.24)$$

$$C_a = c_a \left(L - \frac{D}{\sin \beta} \right) \quad (5.25)$$

By balancing the forces in the vertical direction, the following formulation
results:

$$E_A \sin \left(\frac{\omega + \beta}{2} \right) = W_A - N_A \cos \beta - \frac{N_A \tan \delta + C_a}{FS} (\sin \beta) \quad (5.26)$$

Hence, the interwedge force acting on the active wedge is:

$$E_A = \frac{(FS)(W_A - N_A \cos \beta) - (N_A \tan \delta + C_a)\sin \beta}{\sin \left(\dfrac{\omega + \beta}{2} \right)(FS)} \quad (5.27)$$

The passive wedge can be considered in a similar manner:

$$W_P = \frac{\gamma}{2 \tan \omega} \left[\left(L - \frac{D}{\sin \beta} - h_c \tan \beta \right) (\sin \beta - \cos \beta \tan \omega) + \frac{h_c}{\cos \beta} \right]^2 \quad (5.28)$$

$$N_p = W_P + E_P \sin \left(\frac{\omega + \beta}{2} \right) \quad (5.29)$$

$$C = \frac{\gamma}{\tan \omega} \left[\left(L - \frac{D}{\sin \beta} - h_c \tan \beta \right) (\sin \beta - \cos \beta \tan \omega) + \frac{h_c}{\cos \beta} \right] \quad (5.30)$$

By balancing the forces in the horizontal direction, the following formulation results:

$$E_P \cos \left(\frac{\omega + \beta}{2} \right) = \frac{C + N_P \tan \phi}{FS} \quad (5.31)$$

Hence, the interwedge force acting on the passive wedge is:

$$E_P = \frac{C + W_P \tan \phi}{\cos \left(\dfrac{\omega + \beta}{2} \right) (FS) - \sin \left(\dfrac{\omega + \beta}{2} \right) \tan \phi} \quad (5.32)$$

Again, by setting $E_A = E_P$, the following equation can be arranged in the form of $ax^2 + bx + c = 0$ which in our case is

$$a(FS)^2 + b(FS) + c = 0 \quad (5.13)$$

where

$$a = (W_A - N_A \cos \beta)\cos \left(\frac{\omega + \beta}{2} \right)$$

$$b = -[(W_A - N_A \cos \beta)\sin \left(\frac{\omega + \beta}{2} \right) \tan \phi$$

$$+ (N_A \tan \delta + C_a)\sin \beta \cos \left(\frac{\omega + \beta}{2} \right)$$

$$+ \sin \left(\frac{\omega + \beta}{2} \right) (C + W_P \tan \phi)]$$

$$c = (N_A \tan \delta + C_a)\sin \beta \sin \left(\frac{\omega + \beta}{2} \right) \tan \phi \quad (5.33)$$

Again, the resulting FS value can then be obtained using equation (5.15).

To illustrate the use of the above developed equations, the design curves of Figure 5.12 are offered. They show that the FS value increases in proportion to greater cover soil thicknesses at the toe of the slope compared to the thickness at the crest. This is evidenced by a shallower finished cover

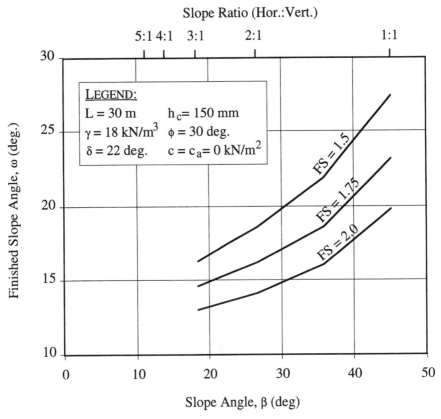

FIG. 5.12. *Design Curves for FS Values for Different Upper and Lower Cover Soil Slope Angles*

soil slope angle than that of the slope angle of the geomembrane and the soil beneath, i.e., the value of "ω" being less than "β." Note that the curves are developed specifically for the variables stated in the legend. Example problem #3 illustrates the use of the curves.

Example 3: Given: a 30 m long slope with a tapered thickness cover soil of 150 mm at the crest extending at an angle "ω" of 16 deg. to the toe. The unit weight of the cover soil is 18 kN/m³. The soil has a friction angle of 30 deg. and zero cohesion, i.e., it is a sand. The interface friction angle with the underlying geomembrane is 22 deg. with zero adhesion. What is the FS value at an underlying soil slope angle "β" of 3(H)-to-1(V), i.e., 18.4 deg.?

Solution: Using the design curves of Figure 5.12 (which were developed for the exact conditions of the example problem), the resulting FS = 1.57.

Comment: The result of this problem (with tapered thickness cover soil) is FS = 1.57, versus example #1 (with a uniform thickness cover soil) which gave FS = 1.25. Thus the increase in FS value is 24%. Note, however, that at ω = 16 deg., the thickness of the cover soil at the toe of the slope is approximately 1.4 m. Thus, the increase in cover soil *volume* used over example #1 is approximately 165%. The trade-offs between these two issues should be considered when using the strategy of tapering cover soil thickness to increase the FS value of a particular cover soil slope.

5.3.4 Veneer Reinforcement of Cover Soil Slopes

A different way of increasing a given slope's factor of safety is to reinforce it with a geosynthetic material. Such reinforcement can be either *intentional* or *non-intentional*. By intentional, we mean to include a geogrid or high strength geotextile within the cover soil to purposely reinforce the system against instability (see Figure 5.13). Depending on the type and

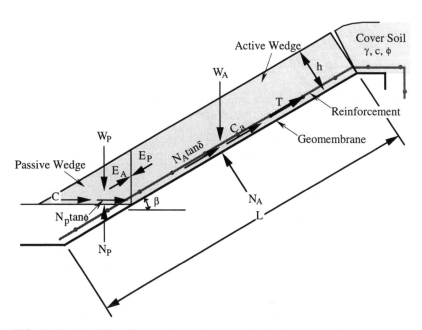

FIG. 5.13. *Limit Equilibrium Forces Involved in Finite Length Slope Analysis Including Use of Veneer Reinforcement*

amount of reinforcement, the majority, or even all, of the gravitational stresses can be supported, resulting in a major increase in the FS value. By non-intentional, we refer to multi-component liner systems where a low shear strength interface is located beneath an overlying geosynthetic(s). In this case, the overlying geosynthetic(s) is inadvertently acting as veneer reinforcement to the system. In some cases, the designer may not realize that such geosynthetics are being stressed in an identical manner as a geogrid or high strength geotextile, but they are. Intentional, or non-intentional, the stability analysis is identical.

As seen in Figure 5.13, the analysis follows Section 5.3.1, but a force from the reinforcement "T," acting parallel to the slope, provides additional stability. This force "T," acts only within the active wedge. By taking free body force diagrams of the active and passive wedges, the following formulation for the factor of safety results. All symbols used in Figure 5.13 were previously defined (see Section 5.3.1) except the following:

$T = T_{reqd}$ = the required (long-term design) strength of the geosynthetic

Considering the active wedge, by balancing the forces in the vertical direction, the following formulation results:

$$E_A \sin \beta = W_A - N_A \cos \beta - \left(\frac{N_A \tan \delta + C_a}{FS} + T \right) \sin \beta \quad (5.34)$$

Hence, the interwedge force acting on the active wedge is:

$$E_A = \frac{(FS)(W_A - N_A \cos \beta - T \sin \beta) - (N_A \tan \delta + C_a)\sin \beta}{\sin \beta(FS)} \quad (5.35)$$

Again, by setting $E_A = E_P$, [see Equation (5.12) for the expression of E_P], the following equation can be arranged in the form of Equation (5.13), in which the "a," "b," and "c" terms are defined as follows:

$a = (W_A - N_A \cos \beta - T \sin \beta)\cos \beta$

$b = -[(W_A - N_A \cos \beta - T \sin \beta)\sin \beta \tan \phi + (N_A \tan \delta + C_a)\sin \beta \cos \beta$

$\quad + \sin \beta(C + W_P \tan \phi)]$

$c = (N_A \tan \delta + C_a)\sin^2\beta \tan \phi \quad (5.36)$

Again, the resulting FS value can be obtained using Equation (5.15).

The value of required, or design, tensile strength is necessary for the long-term stability of the cover soil. In order to obtain an ultimate, or as-manufactured, tensile strength of the geogrid or high strength geotextile, the value of T_{reqd} must be *increased* for site-specific conditions via various factors (all ≥ 1.0). Such values as installation damage, creep, and long-term degradation are generally considered [see Equation (5.37) after Koerner

1994]. Note that if seams are involved in the reinforcement, a reduction factor should be added accordingly.

$$T_{ult} = T_{reqd}(RF_{ID} \times RF_{CR} \times RF_{CBD}) \qquad (5.37)$$

where

T_{reqd} = required, or design, value of reinforcement strength
T_{ult} = ultimate (as-manufactured) value of reinforcement strength
RF_{ID} = factor for installation damage
RF_{CR} = factor for creep
RF_{CBD} = factor for chemical/biological degradation

To illustrate the use of the above-developed equations, the design curves of Figure 5.14 have been drawn. They show the improvement of FS values

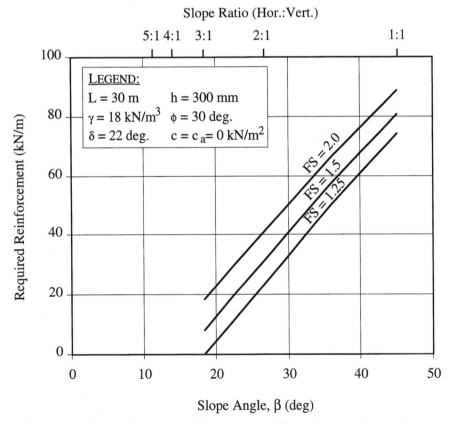

FIG. 5.14.   *Design Curves for FS Values for Different Slope Angles and Veneer Reinforcement Strengths*

with increasing strength of the reinforcement. Note that the curves are developed specifically for the variables stated in the legend. Example problem #4 illustrates the use of the design curves.

Example 4: Given: a 30 m long slope with a uniform thickness cover soil of 300 mm and a unit weight of 18 kN/m^3. The soil has a friction angle of 30 deg. and zero cohesion, i.e., it is a sand. The proposed reinforcement is a geogrid with a design wide-width tensile strength of 10 kN/m. Thus, the reduction factors in equation (5.37) have already been included. The geogrid apertures are large enough that the cover soil will strike through and provide an interface friction angle with the underlying geomembrane of 22 deg. and zero adhesion. What is the FS value at a side slope angle of 3(H)-to-1(V), i.e., 18.4 deg.?

Solution: Using the design curves of Figure 5.14 (which were developed for the exact conditions of the example problem), the resulting FS = 1.57.

Comments: Note that the use of T_{reqd} = 10 kN/m in the analysis will require a T_{ult} value of the geogrid per equation (5.37). For example, if the cumulative reduction factor in equation (5.37) was 4.0, the ultimate (as-manufactured) strength of the geogrid would have to be 40 kN/m. Also, note that this same type of analysis could also be used for high strength geotextile reinforcement. The only difference is that strike-through of the cover soil will not occur. Hence, the geotextile must carry the difference in interface friction values between the cover soil above and the geomembrane below. The analysis follows along the same general lines as presented here.

It should be emphasized that the preceding analysis is focused on *intentionally* improving the FS value by the inclusion of geosynthetic reinforcement. This is provided by geogrids or high strength geotextiles being placed above the upper surface of the low strength interface material. It is often placed directly above a geomembrane or within the cover soil itself.

Interestingly, some amount of veneer reinforcement is often *non-intentionally* provided by a protection geotextile placed beneath the cover soil and above a geomembrane-lined slope. The geotextile will often be highly stressed by the cover soil, and if its interface friction strength against the underlying geomembrane is relatively low, it will tend to slide on the underlying geomembrane. When contained in an anchor trench, however, the protection geotextile actually acts as a de facto reinforcement material. Since its wide-width tensile strength is usually low, it does relatively little to improve the slope's factor of safety. Furthermore, when it fails or pulls out of its anchor trench, the sliding of the geotextile (and overlying cover soil) on the underlying (and stationary) geomembrane is very abrupt. Clearly, if veneer reinforcement is to be provided, it must be done intentionally with the proper type of geosynthetic, and designed accordingly.

5.4 CONSIDERATION OF SEEPAGE FORCES

The previous section presented the general problem of slope stability analysis for final cover soils placed on slope angles of varying degrees. The tacit assumption throughout was that a drainage layer was placed above the barrier layer with adequate capacity to remove, i.e., transmit, permeating water parallel to the slope and safely away from the cross section. The amount of water to be removed is obviously a site-specific situation. Note that in arid areas, drainage may not be required.

Unfortunately, adequate drainage of final covers has sometimes not been available, and seepage-induced slope stability problems have occurred. The following situations have resulted in seepage-induced slides:

- Inadequate drainage capacity at the toe of the slope, where seepage quantities accumulate and are at their maximum.
- Fines from quarried stone accumulating at the toe of the slope, thereby decreasing the as-constructed permeability over time.
- Fine, cohesionless, cover soil particles migrating through the filter (if one is even present) and the drainage layer, and then accumulating at the toe of the slope, thereby decreasing the as-constructed permeability over time.
- Freezing of the drainage layer at the toe of the slope, while the top of the slope thaws, thereby mobilizing seepage forces against the ice wedge at the toe.

These types of seepage-induced failures appear to require a variation in slope stability design methodology, which is the focus of this subsection. Detailed discussion is given in Soong and Koerner (1996).

Consider a cover soil of uniform thickness placed directly above a geomembrane at a slope angle of "β" as shown Figure 5.15. Unlike previous examples, however, a saturated soil zone exists within the cover soil for part or all of the thickness. The saturated boundary is shown as two different phreatic surface orientations. This is because seepage can be built up in the cover soil in two different ways: a horizontal buildup from the toe and a parallel-to-slope buildup. These two hypotheses are defined and quantified as a horizontal submergence ratio (HSR) and a parallel submergence ratio (PSR). The dimensional definitions of both ratios are given in Figure 5.15.

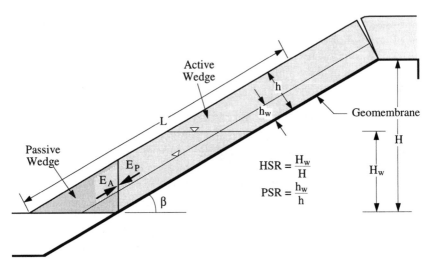

FIG. 5.15. Cross Section of Cover Soil on a Geomembrane Illustrating Different Submergence Assumptions and Related Definitions

When analyzing the stability of slopes using the limit equilibrium method, free body diagrams of the passive and active wedges are taken with the appropriate forces (now including pore water pressures) being applied. Note that the two interwedge forces, namely, E_a and E_p, are also shown in Figure 5.15. The formulation for the resulting factor of safety, for horizontal seepage buildup and then for parallel to slope seepage buildup, follows.

The Case of Horizontal Seepage Buildup.   Figure 5.16 shows the free body diagram of both the active and the passive wedges, assuming horizontal seepage buildup.

All symbols used in Figure 5.16 were previously defined except the following:

$\gamma_{sat'd}$ = saturated unit weight of the cover soil
γ_{dry} = dry unit weight of the cover soil
γ_W = unit weight of water
H = vertical height of the slope measured from the toe
H_W = vertical height of the free water surface measured from the toe
U_h = resultant of the horizontal pore pressures acting on both wedges
U_n = resultant of the pore pressures acting perpendicular to the slope

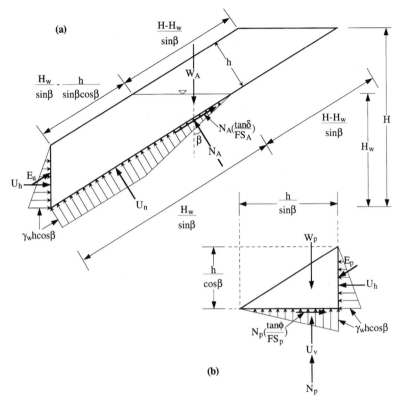

FIG. 5.16. *Free Body Diagrams of Cover Soil with Horizontal Seepage Buildup*

U_v = resultant of the vertical pore pressures acting on the passive wedge

The expression for finding the factor-of-safety can be derived as follows:

Considering the active wedge,

$$W_A = \left(\frac{\gamma_{sat'd}(h)(2H_w \cos\beta - h)}{\sin 2\beta}\right) + \left(\frac{\gamma_{dry}(h)(H - H_w)}{\sin\beta}\right) \quad (5.38)$$

$$U_n = \frac{\gamma_w(h)(\cos\beta)(2H_w \cos\beta - h)}{\sin 2\beta} \quad (5.39)$$

$$U_h = \frac{\gamma_w h^2}{2} \quad (5.40)$$

$$N_A = W_A(\cos\beta) + U_h(\sin\beta) - U_n \quad (5.41)$$

The interwedge force acting on the active wedge can then be expressed as:

$$E_A = W_A \sin \beta - U_h \cos \beta - \frac{N_A \tan \delta}{FS} \tag{5.42}$$

The passive wedge can be considered in a similar manner and the following expressions result:

$$W_P = \frac{\gamma_{sat'd} h^2}{\sin 2\beta} \tag{5.43}$$

$$U_V = U_h \cot \beta \tag{5.44}$$

The interwedge force acting on the passive wedge can then be expressed as:

$$E_P = \frac{U_h(FS) - (W_P - U_V)\tan \phi}{\sin \beta \tan \phi - \cos \beta(FS)} \tag{5.45}$$

Again, by setting $E_A = E_P$, the following equation can be arranged in the form of $ax^2 + bx + c = 0$ which in this case is

$$a(FS)^2 + b(FS) + c = 0 \tag{5.13}$$

where

$$a = W_A \sin \beta \cos \beta - U_h \cos^2\beta + U_h$$
$$b = -W_A \sin^2\beta \tan \phi + U_h \sin \beta \cos \beta \tan \phi$$
$$\quad - N_A \cos \beta \tan \delta - (W_P - U_V)\tan \phi$$
$$c = N_A \sin \beta \tan \delta \tan \phi \tag{5.46}$$

Again, the resulting FS value is obtained using Equation (5.15).

The Case of Parallel-to-Slope Seepage Buildup.   Figure 5.17 shows the free body diagrams of both the active and passive wedges with seepage buildup in the direction parallel to the slope. Identical symbols as defined in the previous case are used here with an additional definition of h_W equal to the height of free water surface measured in the direction perpendicular to the slope.

Note that the general expression of factor of safety shown in Equation (5.15) is still valid. However, the a, b, and c terms shown in Equation (5.46) have different definitions in this case owing to the new definitions of the following terms:

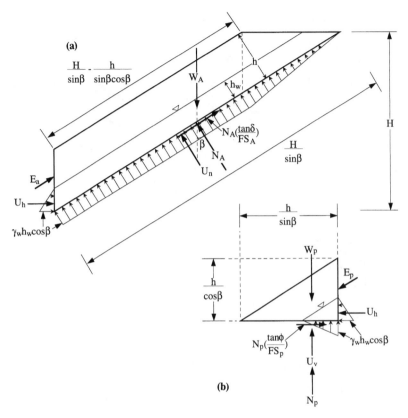

FIG. 5.17. *Free Body Diagrams of Cover Soil with Parallel-to-Slope Seepage Buildup*

$$W_A = \frac{\gamma_{dry}(h - h_w)(2H \cos \beta - (h + h_w)) + \gamma_{sat'd}(h_w)(2H \cos \beta - h_w)}{\sin 2\beta} \quad (5.47)$$

$$U_n = \frac{\gamma_w h_w \cos \beta(2H \cos \beta - h_w)}{\sin 2\beta} \quad (5.48)$$

$$U_h = \frac{\gamma_w(h_w)^2}{2} \quad (5.49)$$

$$W_P = \frac{\gamma_{dry}(h^2 - h_w^2) + \gamma_{sat'd}(h_w^2)}{\sin 2\beta} \quad (5.50)$$

In order to illustrate the effect of the above developed equations, the design curves of Figure 5.18 have been drawn. They show the decrease in FS value with increasing submergence ratio for all values of interface fric-

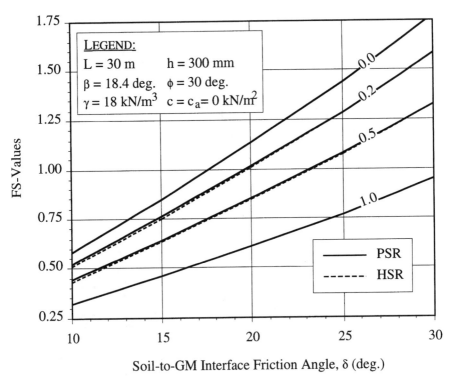

FIG. 5.18.   *Design Curves for Stability of Cover Soils for Different Submergence Ratios*

tion. Furthermore, the differences in response curves for the parallel and horizontal submergence ratio assumptions are seen to be very small. Note that the curves are developed specifically for the variables stated in the legend. Example problem #5 illustrates the use of the design curves.

Example 5: Given: a 30 m long slope with a uniform thickness cover soil of 300 mm at a dry unit weight of 18 kN/m³. The soil has a friction angle of 30 deg. and zero cohesion, i.e., it is a sand. The soil becomes saturated through 50% of its thickness, i.e., it is a parallel seepage problem with PSR = 0.5, and its saturated unit weight increases to 21 kN/m³. Direct shear testing has resulted in an interface friction angle of 22 deg. and zero adhesion. What is the factor of safety at a slope angle of 3(H)-to-1(V), i.e., 18.4 deg.?

Solution: Using the design curves of Figure 5.18 (which were developed for the exact conditions of the example problem), the resulting FS = 0.93.

Comment: The seriousness of seepage forces in a slope of this type is immediately obvious. Had the saturation been 100% of the thickness, the FS value would have been even lower. Furthermore, the result from a horizontal assumption of saturated cover soil with the same saturation ratio will give almost identically low FS values. Clearly, the lesson to take from this example problem is that adequate *long-term* drainage above the barrier layer in cover soil slopes must be provided to avoid saturation from occurring.

5.5 CONSIDERATION OF SEISMIC FORCES

In areas of anticipated earthquake activity, the slope stability analysis of a final cover soil over an engineered landfill, abandoned dump, or remediated site must consider seismic forces. Subtitle "D" of the United States EPA regulations requires such an analysis for sites that have experienced $0.1g$ horizontal acceleration, or more, within the past 250 years. As seen in Figure 5.19, this includes not only the western areas in the United States, but major sections in the midwest and northeast areas as well. If practiced worldwide, such a criterion would have huge implications.

Seismic analysis of cover soils of the type under consideration in this book is a two-part process:

- The calculation of a FS value using a pseudo-static analysis with the addition of a horizontal force acting at the centroid of the cover soil cross section.
- If the FS value in the above calculation falls below 1.0, a permanent deformation analysis is required. The calculated deformation is then assessed in light of the potential damage to the cover soil section and either accepted, or the slope requires an appropriate redesign. The redesign is then analyzed until the situation becomes acceptable.

The **first part** of the analysis is a pseudo-static approach which follows the previous examples, except for the addition of a horizontal force at the centroid of the cover soil in proportion to the anticipated seismic activity. Figure 5.19 gives average seismic coefficients for various zones in the United States. Similar maps are available on a worldwide basis. Note that C_s is nondimensional and is a ratio of the bedrock acceleration to gravitational acceleration. This value of C_s is modified using available computer codes such as "SHAKE" (see Schnabel et al. 1972), for propagation to the

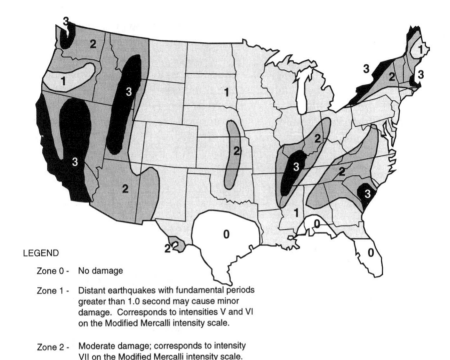

LEGEND

Zone 0 - No damage

Zone 1 - Distant earthquakes with fundamental periods
greater than 1.0 second may cause minor
damage. Corresponds to intensities V and VI
on the Modified Mercalli intensity scale.

Zone 2 - Moderate damage; corresponds to intensity
VII on the Modified Mercalli intensity scale.

Zone 3 - Major damage; corresponds to intensity VIII
and higher on the Modified Mercalli intensity scale.

Seismic Coefficients Corresponding to Each Zone.

Zone	Intensity of Modified Mercalli Scale	Average Seismic Coefficient (C_s)	Remark
0	—	0	No damage
1	V and VI	0.03 to 0.07	Minor damage
2	VII	0.13	Moderate damage
3	VII and higher	0.27	Major damage

FIG. 5.19. Seismic Zone Map and Related Seismic Coefficients for Continental U.S. [after Algermissen (1960)]

landfill cover as shown in Figure 5.20. In most cases it will be amplified. For detailed discussion, see Seed and Idriss (1982) and Idriss (1990). The analysis then proceeds as follows.

All symbols used in Figure 5.20 have been previously defined and the expression for finding the factor of safety can be derived as follows:

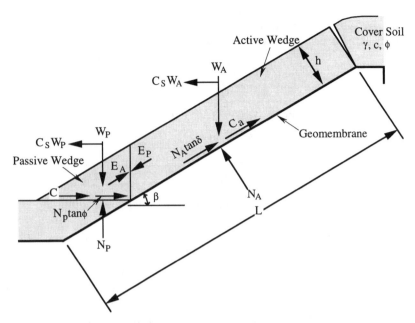

FIG. 5.20. *Limit Equilibrium Forces Involved in Pseudo-Static Analysis Using Average Seismic Coefficient*

Considering the active wedge, by balancing the forces in the horizontal direction, the following formulation results:

$$E_A \cos \beta + \frac{(N_A \tan \delta + C_a)\cos \beta}{FS} = C_s W_A + N_A \sin \beta \qquad (5.51)$$

Hence, the interwedge force acting on the active wedge results:

$$E_a = \frac{(FS)(C_s W_A + N_A \sin \beta) - (N_A \tan \delta + C_a)\cos \beta}{(FS)\cos \beta} \qquad (5.52)$$

The passive wedge can be considered in a similar manner and the following formulation results:

$$E_P \cos \beta + C_s W_P = \frac{C + N_P \tan \phi}{FS} \qquad (5.53)$$

Hence, the interwedge force acting on the passive wedge is:

$$E_P = \frac{C + W_P \tan \phi - C_s W_P(FS)}{(FS)\cos \beta - \sin \beta \tan \phi} \qquad (5.54)$$

Again, by setting $E_A = E_P$, the following equation can be arranged in the form of $ax^2 + bx + c = 0$ which in this case is

$$a(FS)^2 + b(FS) + c = 0 \qquad (5.13)$$

where

$a = (C_SW_A + N_A \sin \beta)\cos \beta + C_SW_P \cos \beta$

$b = -[(C_SW_A + N_A \sin \beta)\sin \beta \tan \phi + (N_A \tan \delta + C_a)\cos^2\beta$
$\quad + (C + W_P \tan \phi)\cos \beta]$

$c = (N_A \tan \delta + C_a)\cos \beta \sin \beta \tan \phi \qquad (5.55)$

The resulting FS value is obtained from the following equation:

$$FS = \frac{-b + \sqrt{b^2 - 4ac}}{2a} \qquad (5.15)$$

Using these concepts, a typical design curve for various FS values, as a function of seismic coefficient, can be developed (see Figure 5.21). Note that the curve is developed specifically for the variables stated in the legend. Example problem 6(a) illustrates the use of the curve.

Example 6(a): Given: a 30 m long slope with uniform thickness cover soil of 300 mm at a unit weight of 18 kN/m³. The soil has a friction angle of 30 deg. and zero cohesion, i.e., it is a sand. The cover soil is on a geomembrane, as shown in Figure 5.20. Direct shear testing has resulted in an interface friction angle of 22 deg. with zero adhesion. The slope angle is 3(H)-to-1(V), i.e., 18.4 deg. A design earthquake at the site results in an average seismic coefficient of 0.10. What is the FS value?

Solution: Using the design curve of Figure 5.21 (which was developed for the exact conditions of the example problem), the resulting FS = 0.94.

Comment: Had the above FS value been greater than 1.0, the analysis would be complete, the assumption being that cover soil stability can withstand the short-term excitation of an earthquake and still not fail. However, since the value is less than 1.0, a second part of the analysis is required.

The **second part** of the analysis is based on calculating the estimated deformation of the lowest shear strength interface in the cross section under consideration. The deformation is then assessed in light of the potential damage that may be imposed on the final cover system.

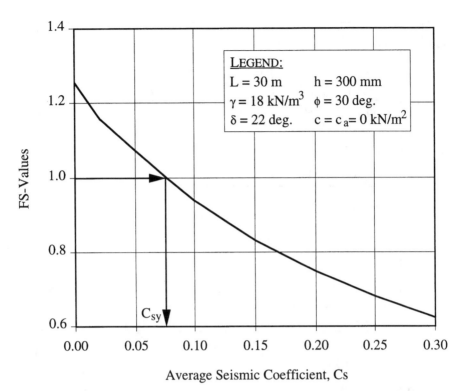

FIG. 5.21. *Design Curves for Stability Using Average Seismic Coefficient in Pseudo-Static Seismic Analysis*

In a permanent deformation analysis, a record earthquake must be selected for the design process. This is a critical decision, not only from the perspective of the magnitude (spectrum), but also from the perspective of its location with respect to the site under consideration. Once selected (and agreed upon by all parties involved), a computer model is used to (a) transfer the recorded spectrum horizontally to the actual site location, and (b) bring the transferred spectrum vertically through the overburden soil and solid waste to the final cover. This is done using a computer model, a number of which are commercially available, e.g., "SHAKE" is commonly used in the United States.

To begin the permanent deformation analysis, a yield acceleration, "C_{sy}," is obtained from a pseudo-static analysis under an assumed FS = 1.0. Figure 5.21 illustrates this procedure for the assumptions stated in the legend. It results in a value of C_{sy} = 0.075. Coupling this value with the spectrum obtained for the actual site location and cross section results in a comparison, as shown in Figure 5.22a. If the earthquake spectrum never

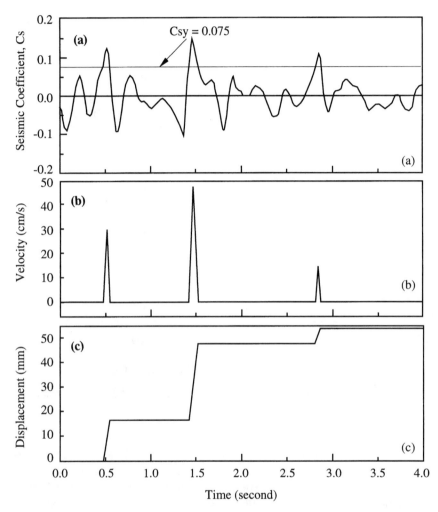

FIG. 5.22.   Design Curves to Obtain Permanent Deformation Using: (a) Acceleration; (b) Velocity; (c) Displacement Curves

exceeds the value of C_{sy}, there is no anticipated permanent deformation. However, whenever any part of the spectrum exceeds the value of C_{sy}, permanent deformation is expected. By double integration of the acceleration spectrum, to velocity and then to displacement, the anticipated value of deformation can be obtained. This value is considered to be permanent deformation, and is then assessed based on the site-specific implications of damage to the final cover system. Example 6(b) continues the previous pseudo-static analysis into the deformation calculation.

Example 6(b): Continue the example problem #6(a) and determine the anticipated permanent deformation of the weakest interface in the cover soil system. The site-specific design spectrum is given in Figure 5.22a.

Solution: The interface of concern is the cover soil-to-geomembrane surface for this particular example. With a yield acceleration of 0.075 from Figure 5.21 and the site-specific design spectrum shown in Figure 22a, double integration produces Figures 22b and c. The three peaks, exceeding the yield acceleration value of 0.075, produce a permanent deformation of approximately 54 mm. This value is now viewed in light of the deformation capability of the cover soil above the particular geomembrane used at the site.

Comments: An assessment of the implications of deformation (in this example it is 54 mm) is very subjective. For example, this problem could easily have been framed to produce much higher permanent deformations. Such deformations can readily be envisioned in high earthquake-prone areas. At the minimum in such an assessment of cover soil systems, the concerns for appurtenances and ancillary piping must be addressed.

5.6 SUMMARY

This chapter has focused on the mechanics of analyzing slopes as part of final cover systems on engineered landfills, abandoned dumps, and re-mediated waste piles. Design curves in all of the sections have resulted in global FS values. Each section was presented from a designer's perspective in moving from the simplest to the most advanced. Table 5.1 summarizes

TABLE 5.1. Summary of Slope Stability Analyses for Similar Conditions, per Example Problems in this Section

Example	Condition	FS-value
1	standard example*	1.25
2(a)	equipment up-slope	1.24
2(b)	equipment down-slope	1.03
3	tapered thickness	1.57
4	veneer reinforcement	1.57
5	seepage	0.93
6	seismic	0.94

*30 m long slope, sand cover soil of 300 mm thickness, ϕ = 30 deg., on a geomembrane with δ = 22 deg., at a slope angle β = 18.4 deg.

the factors of safety of the similarly framed example problems so that insight can be gained from each of the conditions analyzed. Throughout the chapter, the inherent danger of building a relatively steep slope on a potentially weak interface material, oriented in the exact direction of a potential slide, should have been apparent. The standard example was purposely made to have a low, but not at failure, factor of safety. The FS value was seen to decrease with construction equipment on the slope, particularly when moving down-slope. This situation should be avoided. While the techniques of a tapered cover soil thickness and veneer reinforcement using geosynthetics both result in significant improvement of the slope's stability, both have implications; the former being additional soil cost and toe space requirements, the latter being the cost of the geosynthetics.

Other options available to the designer to increase the stability of a given situation are as follows:

1. Decrease the slope angle either uniformly by a constant amount, or gradually from the toe to the crest, i.e., crown the cover soil topography into a convex shape.
2. Interrupt long slope lengths with intermediate benches, or even berms, if erosion conditions so warrant.
3. Provide for discrete zones of the cover surface topography, each with its separate grading pattern. Thus one will have an accordion folded type of grading scheme, instead of a uniform final topography.
4. Provide for a temporary cover until the majority of the subsidence occurs in the waste mass (perhaps waiting for 5 to 10 years), which has the effect of decreasing the toe-to-crest elevation difference. Construct the final cover at that time.
5. Work with lower than typical factors of safety during the postclosure period. If a slide occurs, design the cross section so that the barrier layer is not effected, and then repair the slope accordingly.

Lastly, the decrease in FS value from seepage forces and from seismicity were both illustrated. Clearly, seepage forces should be avoided by providing proper drainage above the barrier layer. On the other hand, seismicity is site-specific and must be dealt with accordingly. Recommendations are given in Soong and Koerner (1996).

In this latter context, we conclude with a discussion on factor of safety (FS) values. Note that we are referring to the global FS value, not reduction factors which necessarily must be placed on geosynthetic reinforcement materials when they are present. In general, one can consider global FS values to vary in accordance with the site-specific issue of required service

TABLE 5.2. Qualitative Rankings for Global Factor of Safety
Values in Performing Stability Analyses of Final Cover Systems

Duration → ↓ Concern	Temporary	Permanent
Noncritical	Low	Moderate
Critical	Moderate	High

time (i.e., duration) and the implication of a slope failure (i.e., the concern).
Table 5.2 gives the general concept in qualitative terms. Using the above
as a conceptual guide, the authors recommend the use of the minimum
global factor of safety values listed in Table 5.3, as a function of the type
of underlying waste.

A "low" ranking refers to a relatively non-critical structure, where the
risk of a failure is accepted and the consequences are minimal (e.g., no
environmental impact and no risk of harm to people). An example of a low
ranking might be a small, temporary slope. A medium ranking refers to a
riskier situation, for instance, where there might be some environmental
impact (but not severe impact). Clearly the most important factor in establishing rankings is the potential for harm to people. Whenever a slope failure endangers people, the ranking should be high. If other consequences
of failure are severe (e.g., environmental impact), the ranking should also
be high. Higher factors of safety are recommended for hazardous waste and
abandoned dumps simply because these wastes typically pose the greatest
health threat. However, if sliding could lead to loss of life (for instance,
people living next to the toe of a long, high landfill cover slope), we recommend using a high ranking and the highest factor of safety in Table 5.3
(i.e., FS = 1.6) regardless of the type of waste.

TABLE 5.3. Recommended Global Factor of Safety Values in Performing Stability
Analyses of Final Cover Systems

Type of Waste → ↓ Ranking	Hazardous Waste	Nonhazardous Waste	Abandoned Dumps	Remediated Waste Piles
Low	1.4	1.3	1.4	1.2
Moderate	1.5	1.4	1.5	1.3
High	1.6	1.5	1.6	1.4

It is hoped that the above values give reasonable guidance in final cover slope stability decisions, but it should be emphasized that regulatory agreement and engineering judgment is needed in all situations. We have tried to identify typical issues and problems, but others that we did not discuss are bound to arise from time to time. It is the responsibility of the design engineer to identify all relevant issues, even if they go beyond the scope of coverage in this chapter.

5.7 REFERENCES

Algermissen, S. T. (1969). "Seismic Risk Studies in the United States," Proc. 4th World Conference on Earthquake Engineering, Vol. 1, Santiago, Chile, pp. A1-14 to 27.

Daniel, D. E., Shan, H.-Y., and Anderson, J. D. (1993). "Effects of Partial Wetting on the Performance of the Bentonite Component of a Geosynthetic Clay Liner," Proc. Geosynthetics '93, IFAI Publ., pp. 1483–1496.

Giroud, J. P., and Beech, J. F. (1989). "Stability of Soil Layers on Geosynthetic Lining Systems," Geosynthetics '89 Proceedings, IFAI Publ., pp. 35–46.

Idriss, I. M. (1990). "Response of Soft Soil Sites During Earthquakes," Proc. Symposium to Honor Professor H. B. Seed, Univ. of California, Berkeley, CA.

Koerner, R. M. (1994). *Designing with Geosynthetics*, 3rd Ed., Prentice Hall Book Co., Englewood Cliffs, NJ, 783 pgs.

Koerner, R. M., and Hwu, B.-L. (1991). "Stability and Tension Considerations Regarding Cover Soils on Geomembrane Lined Slopes," Jour. of Geotextiles and Geomembranes, Vol. 10, No. 4, pp. 335–355.

McKelvey, J. A., and Deutsch, W. L. (1991). "The Effect of Equipment Loading and Tapered Cover Soil Layers on Geosynthetic Lined Landfill Slopes," Proceedings of the 14th Annual Madison Waste Conference, Madison, WI, University of Wisconsin, pp. 395–411.

Poulos, H. G., and Davis, E. H. (1974). *Elastic Solutions for Soil and Rock Mechanics*, J. Wiley & Sons, Inc., New York, NY, 411 pgs.

Schnabel, P. B., Lysmer, J., and Seed, H. B. (1972). "SHAKE: A Computer Programs for Earthquake Response Analysis of Horizontally Layered Sites." Report No. EERC 72-12, Earthquake Engineering Research Center, University of California, Berkeley, CA.

Seed, H. B., and Idriss, I. M. (1982). "Ground Motions and Soil Liquefaction During Earthquakes," Monograph No. 5, Earthquake Engineering Research Center, University of California, Berkeley, CA, 134 pgs.

Soong, T.-Y., and Koerner, R. M. (1996). "Seepage Induced Slope Instability," Jour. of Geotextiles and Geomembranes, Vol. 14, No. 7/8, pp. 425–445.

Stark, T. D., and Poeppel, A. R. (1994). "Landfill Liner Interface Strengths from Torsional Ring Stress Tests," Jour. of Geotechnical Engineering, ASCE, Vol. 120, No. 3, pp. 597–617.

Wilson-Fahmy, R. F., and Koerner, R. M. (1993). "Finite Element Analysis of Stability of Cover Soil on Geomembrane Lined Slopes," Proc. Geosynthetics '93, Vancouver, B.C.: IFAI Publ., pp. 1425–1437.

CHAPTER 6

RELATED DESIGNS AND EMERGING CONCEPTS

Prior chapters have dealt with issues common to most engineered landfills, abandoned dumps, and remediated waste sites. This chapter presents some of the more unusual, or esoteric, aspects of solid waste final covers that the authors have been involved with personally, or appear to be provocative as presented in the published literature.

6.1 GENERAL ISSUES

To our knowledge, all federal agencies allow for regulatory variances on final cover design based on the general concept referred to as *technical equivalency*. Technical equivalency implies that if a substitute design, material, component, or system is presented to the permitting agency as being technically equal (or superior) to the regulatory required (or guidance) design, material, component or system, it should be accepted as an alternate. While this concept is intriguing in many situations, there are at least two major hurdles to overcome:

- The regulations rarely, if ever, present the idiosyncrasies (or even a broad template) of specifically how technical equivalency is to be assessed.
- The regulations are usually implemented by states, provinces, or governmental units beneath the federal level wherein the original regulations or guidance had been developed. Regulatory personnel doing the actual implementation often do not feel as comfortable with substitutions based on technical equivalency assessments as did those who drafted the regulations to begin with, and many operate out of fear of setting a bad precedent. Regulatory officials know that once they have accepted an equivalency demonstration for one facility, they must accept similar demonstrations for other facilities. Quite naturally, this

causes regulatory officials to evaluate equivalency demonstrations very cautiously.

In spite of these hurdles, new designs, materials, concepts, and systems have appeared on a regular basis, have been accepted by the permitting agencies, and have been successful in their performance to date. Nowhere is this more apparent than with geosynthetic clay liners (GCLs). The first use of GCLs in waste containment was in 1986, when they appeared as the lower component of a primary liner beneath the waste, to attenuate leakage through the overlying geomembrane. Experience to date has indicated that they have performed as intended (Bonaparte et al. 1996). Furthermore, a technical equivalency template for liners and covers has recently been developed to compare GCLs to compacted clay liners (CCLs) (Koerner and Daniel 1994). It is widely used for technical equivalency purposes for these particular materials.

Inasmuch as the technical equivalency process may appear to be slow and difficult, the proper idea, presented in a rational and open manner, can result in regulatory acceptability. There is every reason to believe that such acceptability will continue.

6.2 OTHER DESIGNS

The general cross sections presented in Chapter 3 emphasize barrier layers that inhibit the infiltration of water through the cover and into the underlying waste. Their assessment is based on saturated conditions of the soil components involved in the barrier layer, as well as in the overlying surface and protection soil layers. This type of design may not be optimal in arid or semi-arid areas, because soil saturation is never achieved.

6.2.1 Capillary Barrier Concept

Work is ongoing at Sandia National Laboratories to evaluate alternative designs (Dwyer 1995). These alternative designs, shown in Figure 6.1, are intended to emphasize the following issues:

- the unsaturated hydraulic conductivity of the soil components,
- an increased water storage potential to allow for eventual evaporation,
- an increased transpiration through engineered vegetative covers, and
- the need to take advantage of local materials for ease of construction and/or substantial cost savings.

Inherent in many arid and semi-arid final cover designs is the inclusion of a capillary barrier in place of a hydraulic barrier (see Figure 6.1a). This

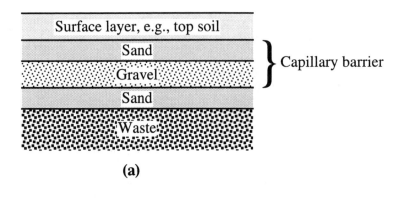

(a)

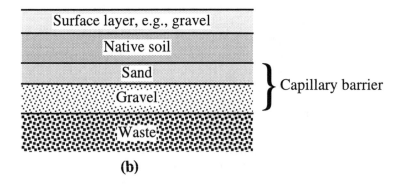

(b)

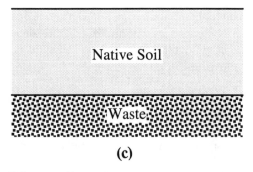

(c)

FIG. 6.1.   Alternative Final Cover Designs in Arid or Semiarid Climate
[after Dwyer (1996)]

concept has been discussed previously. Closely related is a so-called anisotropic barrier (see Figure 6.1b). In both cases, a finer grained soil layer
is placed over a coarser grained soil layer. The difference between the two
concepts lies in where the respective components are located. The capillary
barrier has the materials directly beneath the surface layer, while in the
anisotropic barrier, it is beneath a native soil layer. In both cases, consideration should be given to the use of a geotextile separator placed between
sand and gravel layers. For the enhanced evapotranspiration soil cover, as
shown in Figure 6.1c, the entire cross section is one (relatively thick) layer
of native soil. Field trials are ongoing.

6.2.2 Leachate Recycling

Noted in Section 1.3.2 was the possibility and growing tendency to
reintroduce removed leachate from the base of the landfill and reintroduce
it onto or into its upper surface. Focus in that section was on the amount
and rate of settlement that the waste would undergo. However, the basic
purpose of the procedure is to transform the landfill into a bioreactor,
thereby hastening the waste's decomposition and reducing the long-term
concerns to the local community and adjacent environment. Thus, the crit-
icisms of creating a "dry tomb," which are often voiced by adversaries of
landfills, is greatly muted.

In practicing leachate recycling so as to create a landfill bioreactor, the
leachate can be reintroduced into the waste by a number of different ways
(Shaw and Carey 1996):

1. Surface Application
    - spraying directly from a tank truck
    - distributing from a truck with an attached spray bar
    - spraying from a manifolded pipeline
    - traditional spray irrigation
2. Gravity Infiltration
    - from open trenches
    - from leach beds
    - from vertical wells
3. Active Injection Systems
    - from a manifold system
    - from vertical wells

The latter two methods are illustrated in Figure 6.2. Pohland (1996) presents
the fundamental reaction principles as well as suggestions concerning the
relevant design issues and operational details. The entire proceedings of a
United States EPA (1996) seminar on the procedure provides a wealth of
information.

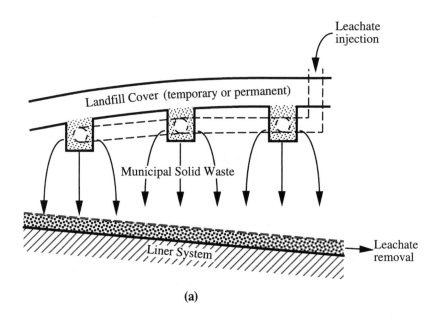

(a)

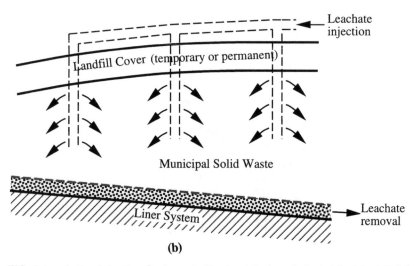

(b)

FIG. 6.2.   *Active Injection Systems to Recirculate Leachate in Municipal Solid Waste Landfills*

6.2.3 Mine Residuals vis-a-vis Land Reclamation

Precious and base metal (hardrock) residuals represent an interesting paradox in the context of this book on final covers for engineered landfills, abandoned dumps, and remediated waste sites, i.e., they are not included. More specifically, the residuals from deep mining or strip mining are generally designated as *ores* and, as such, are excluded from the regulations outlined in Chapter 1. Yet, there are concerns over such residual materials. Miller (1996) lists the following open pit mining concerns:

- Open pit mining often extends below the static groundwater elevation requiring mine dewatering and water management. At closure, the open pit may fill partially or completely with ground and surface water.
- Open pit mining results in relatively steep and high walls. Long-term stability of the walls and public safety are issues for post-closure pits.
- The aesthetic impacts of the open pit are increasingly being identified as an issue of public concern.
- Exposure of acid generating materials in rock walls and floors and the potential for poor water quality in the pit lake or pit lake outflow is a concern.
- Future land use of open pits is often limited, due to safety concerns and an inability to revegetate or stabilize steep walls.

The following are concerns for underground mining:

- Underground mining often extends below the static groundwater elevation requiring mine dewatering and water management. At closure the underground mine may fill partially or completely with groundwater.
- Underground mining results in development of voids, which can cause long-term stability and subsidence issues.
- Discharge water from underground mining operations at closure is considered to be process water by the regulatory agencies and requires a discharge permit and long-term maintenance.

Taken collectively, operations at both open pit and underground mines should be undertaken cognizant of final closures as described herein. Unanticipated closure costs at the end of a mining project will clearly have a negative impact on the overall project economy. Such costs, from proper land reclamation, should be incorporated in the initial mine design and its operations throughout its active life. Furthermore, an ongoing update and re-evaluation of environmental concerns and potential liabilities seems to be a prudent approach for mine owners and operators.

6.3 EMERGING MATERIALS

The multilayered cross sections that were shown in Chapter 3, with each component having its own function(s), present a number of provocative candidates for replacement materials. Indeed, such potential replacements are being regularly assessed and are appearing. In the subsections to follow, they are presented according to their relative positions in a final cover cross section, i.e., from the surface layer downward.

6.3.1 Natural Erosion Control Materials

Clearly, natural vegetation is the backbone of any erosion control system for the surface layer of a final cover in humid or relatively humid locations. Much has been written on the subject, e.g., see the Virginia Erosion and Sediment Control Handbook (1992), from the point of view of a local region. They categorize natural vegetative stabilization as follows:

- Annual Grasses
    - winter rye
    - annual ryegrass
    - German millet
    - Sudangrass
    - annual Lespedeza
- Perennial Grasses
    - tall fescue
    - Kentucky bluegrass
    - redtop
    - Bermudagrass
    - Centipedegrass
- Legumes
    - crown vetch
    - sericea Lespedeza

Beyond this traditional approach, however, there is a resurgence of using more hardy and deeper-rooted vegetation, as illustrated in Figure 6.3. The book by Gray and Sotir (1996), is particularly illuminating in this regard.

6.3.2 Geosynthetic Erosion Control Materials

For many final covers, the establishment of plant species is greatly aided by placing a geosynthetic erosion control layer on the surface layer before, during, or after seeding. Geosynthetic erosion control materials have not commonly been used on landfill covers, but we feel that they should be

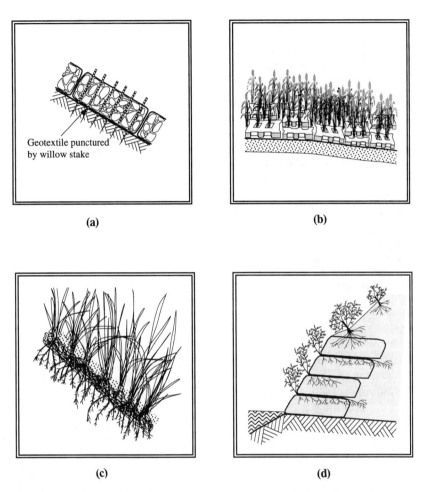

FIG. 6.3. Various Schemes of Erosion Control Using Deep Rooted Natural Vegetation [after PIANC (1996)]

required much more frequently. Major erosion problems are not uncommon. The fact that the completion of construction of covers often occurs at the end of the growing season greatly adds to the risk of erosion. Erosion can be harmful in more ways than simply adding to maintenance costs, e.g., clogging of toe drains by eroded soil. The way to solve these problems is to use and properly install geosynthetic erosion control materials.

The selection of such a material is based upon the slope angle, slope distance, hydrology, time of year, etc. Indeed, there are many such materials to fulfill site-specific needs. Theisen (1992) categorizes the materials as

TABLE 6.1. Geosynthetic Erosion Control Materials [after Theisen (1992)]

(a) Temporary Erosion and Revegetation Materials (TERMS) Straw, hay, and hydraulic mulches Tackifiers and soil stabilizers Hydraulic mulch geofibers Erosion control meshes and nets (ECMNs) Erosion control blankets (ECBs) Fiber roving systems (FRSs)
(b) Permanent Erosion and Revegetation Materials (PERMS)—Soft Armor-Related UV-stabilized fiber roving systems (FRSs) Erosion control revegetation mats (ECRMs) Turf reinforcement mats (TRMs) Discrete length geofibers Vegetated geocellular containment systems (GCSs)
(c) Permanent Erosion and Revegetation Materials (PERMs)—Hard Armor-Related Geocellular containment systems (GCSs) Fabric-formed revetments (FFRs) Vegetated concrete block systems Concrete block systems Stone rip-rap Gabions

shown in Table 6.1. Each category will be described and its typical use illustrated in a final cover application.

6.3.2.1 Temporary Erosion and Revegetation Materials. Temporary erosion and revegetation materials (TERMs) consist of materials which are in whole, or part, degradable. They provide temporary erosion control and are either disposable after a given period, or only function long enougth to facilitate vegetative growth. After the growth is established, the TERM is sacrificed. Some of the products are completely biodegradable, while others are only partially so.

In Table 6.1, the first two products are self-explanatory, consisting of traditional methods of soil erosion control using straw, hay, or mulch loosely bonded by asphalt or adhesive. Their stationability is quite poor. Geofibers, in the form of short pieces of fibers or microgrids, can be mixed into soil with machines or rototillers to aid in laydown and continuity. The fiber or grid inclusions provide for greater stability over straw, hay, or mulch simply broadcast over the ground surface.

Erosion control meshes and nets (ECMNs) are biaxially oriented nets manufactured from polypropylene or polyethylene. They do not absorb moisture, nor do they dimensionally change over time. They are lightweight and are stapled to the previously seeded ground using hooked nails or Ushaped pins. (This is the practice for many of the continuous sheet products to follow.) Stationability is obviously greatly improved over the previously mentioned approaches.

Erosion control blankets (ECBs) are also biaxially oriented nets manufactured from polypropylene or polyethylene, but are placed on one or both sides of a blanket of straw, excelsior, cotton, coconut, or polymer fibers. The fibers are held to the net by glue, lock stitching, or other threading methods.

Fiber roving systems (FRSs) are continuous strands, or yarns, usually of polypropylene, which are fed continuously over the surface to be protected. They can be hand-placed or dispersed using compressed air. After placement on the ground surface, an emulsified asphalt or soil stabilizer is used for controlling position.

TERMs are used in final covers with benches where slopes are relatively flat, e.g., 1 to 2%, as illustrated in Figure 6.4a.

6.3.2.2 Permanent Erosion and Revegetation Materials: Soft Armor-Related.

Within the permanent erosion and revegetation materials (PERMs) is a soft armor-related group, as shown in Table 6.1. These polymer products furnish erosion control, aid in vegetative growth, and eventually become entangled with the vegetation to provide reinforcement to the root system. As long as the material is shielded from sunlight, via shading and soil cover, it will not degrade (at least within the limits of other polymeric materials). The seed is usually applied after the PERM is placed, and is often located directly in the backfilling soil.

The polymers comprising FRSs can be stabilized with carbon black and/or chemical stabilizers such that they can sometimes be considered in the PERM category.

Erosion control revegetation mats (ECRMs) and turf reinforcement mats (TRMs) are closely related to one another, except that ECRMs are placed on the ground surface with a soil infill, while TRMs are placed on the ground surface with soil filling in and above the material. Thus TRMs can be expected to provide better vegetative entanglement and longer performance. Other subtle differences are that ECRMs are usually of greater density and lower mat thickness. Seeding is generally done prior to installation with ECRMs, but is usually done while backfilling within the structure of TRMs.

Discrete length geofibers are short pieces of polymer yarns mixed with soil for the purpose of providing a tensile strength component against sudden forces, as on final covers supporting equipment traffic. Geocellular containment systems (GCSs) consist of three-dimensional cells of geomembrane or geotextile strips, which are filled with soil and, when used for erosion control, are vegetated.

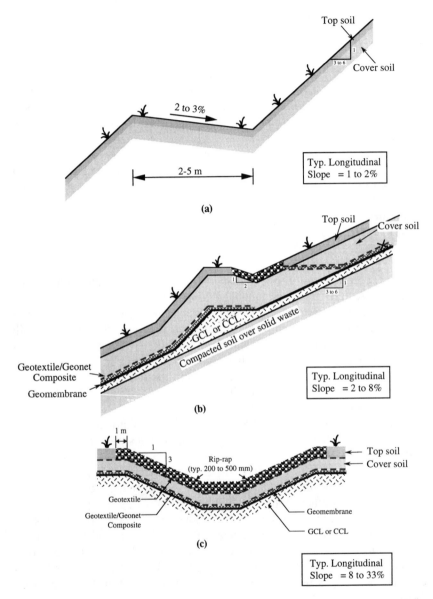

FIG. 6.4. *Various Methods and Cross Sections of Intercepting and/or Channeling Runoff of Final Covers*

Soft armor PERMs are used in final covers with terrace channels where slopes are in the range of 2 to 8%, as illustrated in Figure 6.4b.

6.3.2.3 Permanent Erosion and Revegetation Materials: Hard Armor-Related.
We can include a number of PERMs, which are essentially hard armor systems, in a separate category of inert materials (see Table 6.1).

Whenever the infill material is permanent, as with concrete or grout, GCSs can be considered in this category. Fabric-formed revetments (FFRs) are fabrics sewn together in two layers by mean of drop stitches or gathered in discrete spacings by sewing. The two layers are inflated in their field position by pumping with a flowable grout or concrete mix. When hardened, the surface is either uniform or quilted. The fabric is sacrificial since the hardened grout or concrete provides the hard armor material.

Numerous concrete block systems are available for erosion control. Hand-placed interlocking masonry blocks are also possible. The voids in the blocks and spaces between them are usually vegetated. Alternatively, the system can be factory fabricated as a unit, brought to the job site, and placed on prepared soil. The prefabricated blocks are either laid on, or bonded to, a high strength geotextile substrate. The finished mat can bend and torque by virtue of the blocks being articulated with joints, weaving patterns, or cables. Such systems are generally not vegetated.

Stone rip-rap can be a very effective erosion control method in which large rock is placed on a geotextile substrate. A geotextile placed on the proposed soil surface before rock placement serves as a filter and separator. The stone can vary from small hand-placed pieces to machine-placed pieces of enormous size. Stone rip-rap was described in Chapter 2.

Closely related to rip-rap are gabions, which consist of discrete cells of wire netting filled with hand-placed stone. The wire is usually galvanized steel hexagonal wire mesh, but in some cases can also be a plastic geogrid. Gabions require a geotextile to be placed behind them to act as a filter and separator for the backfilled soil.

Hard armor PERMs are used in final covers with letdown channels, where slopes are in the range of 8 to 33%, as illustrated in Figure 6.4c.

6.3.3 Geofoam

Geofoam is the commonly used term referring to expanded polystyrene (EPS) or extruded polystyrene (XPS) in below-grade applications (Horvath 1994). This lightweight material of a density between 10 and 20 kg/m^3 has unique engineering properties. White (1995) presents the following data as typical of EPS, which is the most commonly used geofoam material:

· water absorption is very low, e.g., 2% by volume at a maximum;

- low temperature, under-water or wet environment, and exposure to freeze-thaw cycling have no negative influence on mechanical properties;
- the insulation value of EPS is excellent, and this feature has been capitalized on in several landfill applications; and
- the mechanical properties of modulus, Poisson's ratio, and strength are readily determined by either static or cyclic loading tests.

Geofoam has been used in final covers above the drainage barrier layer for insulation and for its lightweight properties (Gasper 1990). It has also been used beneath a geomembrane as a smooth protection layer for steep slopes in an abandoned quarry (Horvath 1995).

6.3.4 Shredded Tires

Used automobile and truck tires represent an extremely large quantity of waste material which can be used in waste containment cross-sections. In the United States, the estimated quantity is 200 million waste tires per year. When cut into pieces, ranging from 100 to 250 mm in length, shredded tires have been used as a combination protection layer and drainage layer for leachate collection at the bottom of landfills (Whitty and Ballod 1990). The protection, i.e., cushioning effect, is quite obviously provided by a relatively thick layer of such tire pieces. Not so obvious is the drainage effect, particularly under high normal stresses. Narejo and Shettina (1995) provide permeability data in Figure 6.5.

In the final cover scenario, however, such high normal stresses are generally not encountered. Thus shredded tires could be used as the drainage layer, and if larger quantities are available, for part (or even all) of the protection layer. To the authors knowledge, shredded tires in a final cover have been used on at least one landfill. One caution is expressed, however, as far as the use of tires in final covers is concerned. Over time, tires have been known to rise in elevation from their original positioning. This is a form of solid material fracturing, as opposed to hydrofracturing in soil or rock applications. Tires, or tire pieces, must be maintained stationary in spite of this tendency.

The steel reinforcement in tires is a concern whenever a geomembrane, or other geosynthetic, is directly beneath. In such cases, careful consideration must be given to the design of an adequate protection layer.

6.3.5 Spray-On Elastomeric Liners

Research is ongoing using various types of sprayed elastomers to provide for a hydraulic/gas barrier for final (and temporary) covers. The elastomer is sometimes applied directly on the prepared soil subgrade or (more gen-

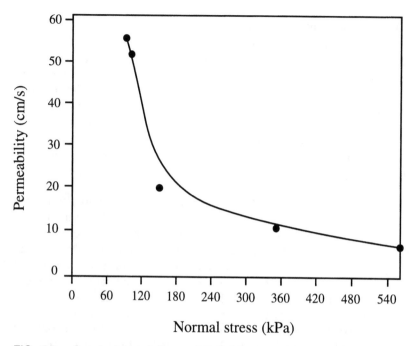

FIG. 6.5. *Constant Head Permeability Test Results According to ASTM D2434 for Rubber Tire Chips of 60 and 100 mm in Size [after Narejo and Shettima (1995)]*

erally) it is sprayed onto a previously placed geotextile substrate. The geotextile is usually a needle-punched nonwoven and it is generally seamed at its edges and ends. It must be laid flat with no wrinkles or folds.

A particular spray elastomer that was evaluated by Cheng et al. (1994) was a polyurea material. Polyurea is an elastomer that is formed by combining two solid components, without the use of a catalyst. The polyurea is applied through a spray process, during which the two components are mixed, heated, and pressurized. The two components used were an amine terminated resin and chain extenders with isocyanates. The chemical formula is shown below, after Primeaux 1991.

$$R{-}NCO + R'NH_2 \rightarrow R{-}NH{-}\overset{\overset{\textstyle O}{\|}}{C}{-}NH{-}R'$$

Isocyanate Amine Polyurea Polymer

Cheng et al. (1994) give results on physical and mechanical properties of this particular elastomer. It should be recognized, however, that there are a tremendous number of possibilities for spray-on liners of this type.

6.3.6 Paper Mill Sludge Liners

Paper mill sludges have been used since 1995 as the barrier layer for some final covers constructed in locations which have such material as a waste product. However, there is little performance data on the engineering properties of the sludges that were used. Moo-Young and Zimmie (1995) performed laboratory tests to determine water content, organic content, specific gravity, permeability, compaction, consolidation, and strength on seven paper mill sludges. They found that the sludges had a high initial water content ranging from 150 to 270 percent, an initial hydraulic conductivity ranging from about 1×10^{-7} cm/s to 5×10^{-6} cm/s, and behaved similarly to highly organic soil.

Moo-Young and Zimmie (1995) also performed laboratory tests on six samples of a sludge used as a barrier layer material. Three samples were obtained shortly after construction and the other three samples were taken at 9, 18, and 24 months after construction. The results of the laboratory tests on these undisturbed samples indicated that the water content and hydraulic conductivity of the sludge decreased somewhat over time as the sludge consolidated and biodegraded (i.e., it mineralized to become more like a soil). The depth of frost penetration in the sludge barrier layer has been monitored since 1992. To date, the frost layer has not penetrated into the sludge layer, due to the protection provided by the overlying soil layers and the high water content of the sludge. Based on their results of laboratory tests over a range of water contents, if a sludge layer is subjected to freezing and thawing cycles, the hydraulic conductivity of the sludge may increase by one to two orders of magnitude.

6.4 EMERGING CONCEPTS AND/OR SYSTEMS

Propelled by the German TA-A regulations requiring that a final cover system for Category III hazardous wastes must have leak-detection capability, a number of in-situ leak monitoring methods are in development. Note that methods of electric leak detection have been used for bottom liners, but they require the barrier system to be flooded with water to carry the current. This is generally not possible for the convex orientation of most final cover barrier systems.

Instead, the general approach has been to place an electrode system on a grid pattern beneath the lower component of the barrier system, in the gas collection layer or the lower part of a compacted clay liner. The grid spacing is obviously an important decision. A close grid pattern, e.g., 1-m spacing, gives excellent leak resolution but is expensive. A wider spaced pattern, e.g., 10-m spacing, gives lower leak resolution but is less expensive. The technique appears to have two different monitoring concepts.

Using stainless steel or copper electrodes, an electrical signal can be transmitted and analyzed based on time-domain reflectometry. The orthogonal signal detection can locate the leak to within the accuracy of the grid spacing pattern. The technique has been available since 1985, but has only been used on a laboratory and experimental basis, see Peggs (1996).

Alternatively, the electrodes can be made from non-metallic conductors, e.g., fiber optics, and analyzed on the basis of optical signal transmission. This is the type of system that is being actively promoted in Germany at the present time (see Rödel 1996).

6.5 TIMING OF CONSTRUCTION OF FINAL COVER FOR MSW LANDFILLS

Final covers for MSW landfills are almost always constructed immediately after the landfill cell is filled to capacity. The result is that a multilayered, sophisticated, expensive cover is placed on a fundamentally unstable waste. Over time, the waste degrades and the cover deforms. Because differential settlement can be severe, the cover often has to be repaired.

In our opinion, a different approach is more rational. We think that rather than constructing the engineered final cover immediately after filling a cell, it would be better, in many situations, to construct a temporary cover that allows controlled infiltration of water (and perhaps leachate recirculation). The temporary cover would be left in place 3 to 10 years, at which time much of the settlement would be complete. At that time, the final cover could be constructed. Because the foundation (underlying MSW) would be more stable, the long-term performance of the final cover would be improved. Current regulations, however, make this suggested practice illegal by requiring prompt construction of the final cover. We urge regulators and landfill owners to rethink the timing of final cover construction.

6.6 FIELD PERFORMANCE OF FINAL COVER SYSTEMS

Some final cover field studies have been conducted in the United States and Europe. They are described in this section, which follows Daniel and Gross (1996).

6.6.1 Cover Failures

There have been a number of documented cases of cover failures, with most of the failures occurring during or shortly after construction and resulting from excessive erosion, build up of seepage pressures in the cover layers, lack of a drainage layer, a drainage layer with insufficient capacity, or an incorrect "estimation" of the shear strength between components within the cross-section. While most of these failures did not involve failure of the barrier layer, they were costly to repair. In one case, severe erosion problems developed because the cover slopes were relatively long (180 m) and the drainage layer was designed with an outlet only at the toe. At some locations, the surface and protection soils had eroded to the top of the clay barrier layer. The erosion problem was exacerbated, in some cases, because the drainage layer outlet at the toe of the slope had not been constructed. In these cases, the trapped water eventually caused pore pressures to become excessive, causing sloughing of the overlying soil layers at the toe of the slope. At another landfill, a gabion-lined channel for surface water slid down the cover slope due to liquefaction of the fine sand beneath it, brought about by high pore-water pressures. It is noted that most of the failures occurred in states with relatively restrictive and prescriptive final cover designs rather than more flexible performance-oriented objectives. This suggests that, in these states with prescriptive final cover designs, greater attention may have been given to regulatory compliance than to the design itself.

6.6.2 Field Study in Hamburg, Germany

Six test covers (10 m wide by 50 m long) were conducted in 1987 to evaluate the field performance of different cover configurations (Melchoir and Miehlich 1989; Melchoir et al. 1994). The covers were constructed with a 750 mm thick sandy loam topsoil layer, underlain by a 250 mm thick fine gravel drainage layer. The drainage layer was underlain by one of four barrier layer types: 1) a 600 mm thick compacted clay layer, 2) a HDPE geomembrane/compacted clay composite layer with welded geomembrane panels, 3) a geomembrane/ compacted clay composite layer with overlapped geomembrane panels, and 4) a compacted clay layer overlying a 600 mm thick fine sand wicking layer and a 250 mm thick coarse sand/fine gravel capillary barrier.

Each of the four cover configurations was constructed on 4 percent or 20 percent slopes, and several configurations were constructed for both slopes. Climate, lateral drainage from the topsoil and drainage layers, runoff, and percolation data are being collected. Soil moisture data are also being collected from several test covers using neutron probes and tensiometers. The preliminary findings of this field study are summarized below.

For the covers with the compacted clay barrier layer, little percolation was observed for the first 20 months after construction. Beginning in August 1989, percolation began to increase and show a correlation with precipitation events. The summer of 1992 was very dry, and tensiometers indicated that the clay layers had undergone more drying than usual. This drying resulting in an almost tenfold increase in percolation measured during the fall of 1992 over percolation recorded a year earlier. Flow through the capillary barrier under one of the compacted clay layers was first observed during this time. When excavations were made into the covers in 1993, the clay layers were found to have small fissures and contain plant roots.

Since 1993, the network of plant roots has developed further, contributing to preferential flow paths and desiccation cracking. Percolation through the compacted clay layers is still increasing and was about 200 mm in 1994.

The covers constructed with a composite barrier layer performed much better, and no percolation has been observed. However, during the summer and fall months, the matrix potential in the clay layers dropped and drainage from the clay layer was recorded. This drainage, which had typically been less than 1.0 mm/yr, has been attributed to thermal gradients. During the summer and early fall, the temperature at the top of the clay layer has been greater than that at the bottom, and water likely flows in liquid and vapor phases from the hotter to cooler regions, resulting in the measured drainage. The water loss caused by thermal gradients has not caused shrinkage of the clay, although the potential for future shrinkage exists.

6.6.3 Field Study in Beltsville, Maryland

Six lysimeters (14 m wide by 21 m long by 3.0 m deep) with 20 percent side slopes were constructed between May 1987 and January 1990 to evaluate final covers incorporating either a compacted clay barrier layer, a rock capillary barrier, or bioengineering management, which combined enhanced runoff and plant transpiration (Schultz et al. 1995). The bioengineering management option used alternating aluminum fiberglass panels as the surface layer over 90 percent of the cover with moisture-stressed vegetation (i.e., Pfitzer junipers) located along gaps in the panels. This latter option requires periodic maintenance and is intended to be used when significant subsidence of the underlying waste is expected. The six lysimeters were constructed with the following covers: (1) bioengineering management, with the initial water level 900 mm above the bottom of the lysimeter; (2) bioengineering management, with the initial water level 1900 mm above the bottom of the lysimeter; (3) reference lysimeter similar to lysimeter 1, except without the surface panels and vegetated with fescue grass; (4) rip-rap surface layer and gravel drainage layer over a compacted clay layer; (4) vegetated soil surface layer, gravel drainage layer, and compacted clay layer over a gravel capillary barrier; and (6) vegetated soil surface layer and gravel drainage layer over a compacted clay layer. All of the covers were constructed over native soil, i.e., not over a waste mass. Rainfall, runoff, deep drainage, and soil moisture content data were collected for the lysimeters.

The data collected through 1994 reveal that initially ponded water in lysimeters 1 and 2 was removed by the plants within two years after construction. The soils in these lysimeters have generally become drier over time. The initial water level in lysimeter 3 rose until it was near the surface and the water had to be pumped out. Deep drainage has been measured for this lysimeter every year of the study. Except for the deep drainage from lysimeter 5, which occurred during 1994, deep drainage has not been observed from lysimeters 4 to 6. It has been noted that the moisture content of the clay layer in lysimeter 4 has been increasing, indicating the possibility of future seepage through the clay. The moisture contents of the clay layers in lysimeters 4 and 6 show seasonal cycling, with the lowest moisture contents being measured in the summer.

6.6.4 Field Study in East Wenatchee, Washington

Two covers (30 m by 30 m), one with a 600 mm thick compacted clay barrier layer overlain by a 150 mm thick vegetated soil surface layer and the other with a 750 mm thick sand capillary barrier layer used in lieu of a compacted clay barrier, were constructed and monitored (Khire 1995). Climate, runoff, percolation, and soil moisture data have been continuously collected since November, 1992. The collected data show that cumulative percolations through the compacted clay and capillary barrier layers have been 31 and 5 mm respectively. Most of the water movement through the capillary barrier occurred in the winter of 1993, primarily due to snowmelt from the relatively high snowfall (1700 mm) occurring that year.

6.6.5 Field Studies in Richland, Washington

Since 1985, Pacific Northwest Laboratories and Westinghouse Hanford Co. have been working to develop a cover design for waste at the Hanford nuclear site. Field tests have been conducted for over the past seven years, using lysimeters to evaluate the performance of different cover materials and configurations (Peterson et al. 1995). Currently, 24 lysimeters are being monitored to assess the effects of varying precipitation, surface soil, and vegetative conditions. No drainage has been measured from lysimeters with vegetated or nonvegetated silt-loam surfaces under normal precipitation conditions; however, some drainage has occurred from lysimeters with nonvegetated silt-loam surfaces under extreme precipitation conditions, e.g., three times normal. Significant quantities of water have drained from lysimeters with gravel and sand surface layers. The performance of one lysimeter with a 150 mm thick layer of nonvegetated silty loam was modeled over a six-year period using the HELP and UNSAT-H models. The HELP model simulation prediction was 1800 percent greater than the observed drainage, while the UNSAT-H model simulation prediction was only 52 percent of the observed drainage.

A prototype cap (0.6 hectare) was aconstructed at the site in 1994 using the following layers, from top to bottom (Gee et al. 1994; Wing and Gee 1994; Peterson et al. 1995):

- 1000 mm thick silt loam/admix gravel surface layer,
- 1000 mm thick silt loam protection layer,

- 150 mm thick sand filter,
- 300 mm thick gravel filter,
- 1500 mm thick fractured basalt rip-rap capillary barrier and biobarrier,
- 300 mm thick gravel cushion and drainage layer,
- 150 mm thick composite asphalt barrier layer,
- 100 mm thick top course, and
- compacted soil foundation.

The composite asphalt barrier layer was discussed previously. Water and wind erosion, biointrusion, revegetation success, and water balance of the cover continue to be monitored. Water balance components being recorded include: precipitation, runoff, snow depth, soil moisture, and percolation. The water balance is being evaluated under normal and stressed (i.e., irrigated) conditions.

6.6.6 Field Study in Idaho Falls, Idaho

A replicate field test program is underway at Idaho National Engineering Laboratories to compare the hydraulic performance of four cover designs: (1) 1000 mm thick vegetated soil layer over a geomembrane/compacted clay composite barrier layer; (2) 2500 mm thick vegetated soil layer with a 500 mm thick biobarrier located within it at 500 mm below the ground surface; (3) the same design as cap 2, except the biobarrier is located 1000 mm below the ground surface; and (4) 2000 mm thick vegetated soil layer. The test plots were constructed in 1993. Two vegetation types have been used, a native mixed plant community and a monoculture of crested wheatgrass. Both vegetative covers were considered, since planted monocultures may be reinvaded by native plant species in the future, and a mixed native plant community may be more resilient to environmental fluctuations. The test plots will, at times, be subjected to burrowing animals, ants, and high levels of irrigation. Climate, soil moisture, and percolation data are being collected. Soil moisture is measured using a neutron probe and time domain reflectometry. Monitoring is ongoing.

6.6.7 Field Studies in Albuquerque, New Mexico

A large-scale field test program is being conducted at Sandia National Laboratories to compare the performance of three covers (each 13 m wide by 100 m long); one incorporates a compacted clay, another a geomembrane/compacted clay composite, and a third has a geomembrane/GCL composite barrier layer. The test plots were constructed and instrumented during 1995 (Dwyer 1995). The hydrology and the erosion of the caps are being monitored. Another three plots were scheduled to be constructed in 1996, each having capillary barriers, with one vegetated to enhance evapotranspiration (recall Figure 6.1). In another field study under limited evapotranspiration conditions, the performances of two gravel capillary barriers were compared (Stormont 1995). When a uniformly-graded sand layer was placed between the gravel barrier and the overlying silty sand, the lateral drainage above the barrier increased while drainage through the gravel decreased.

No data on performance has been collected at this time. Monitoring is expected to continue for the next several years.

6.7 ASSESSMENT OF THE TECHNOLOGY

Several aspects of final cover materials and design, namely the service life of HDPE geomembranes, the internal shear strength of GCLs, the properties of asphalt and paper mill waste, the slope stability of caps, and water balance modeling have been reviewed by Daniel and Gross (1996). Their findings are presented below:

- Preliminary results from long-term laboratory testing of HDPE geomembranes indicate that service life of the geomembranes should be at least several hundred years, depending on the specific product and the site-specific conditions. If HDPE geomembranes provide this length of service, they appear to be a good investment relative to other more costly barrier layer materials, such as compacted clay.
- Based on ongoing field-scale tests of prototype covers incorporating GCLs, the factor of safety against internal shear failure of GCL on 3(H)-to-1(V) slopes has been observed to be 1.5 or greater (Koerner et al. 1996). These results are encouraging, since a cover with a GCL is a more effective barrier against infiltration than a cover with a compacted clay liner. If GCLs remain stable on cover slopes, they appear to be preferable to a compacted clay layer.
- Asphalt and paper mill waste have been used as the barrier layer material for covers. However, the service life of these materials is uncertain. While asphalt may have a longer service life than traditional barrier layer materials, it is significantly more expensive. Also, the significance of flow along asphalt seams is unknown. Paper mill waste involves reuse of a waste product; however, it may be affected by some of the same processes (e.g., freezing/thawing cycles) that adversely impact compacted clay layers.
- The shear strength properties of cover components should be evaluated by conducting project-specific tests under the expected field conditions. The effect of freezing-thawing cycles,

heating-cooling cycles, and creep on the long term shear strength properties should also be considered by using a higher factor of safety or by increasing the thickness of cover soils above the critical layers to provide thermal insulation and isolation from the environment.

· Observations of the seismic performance of several landfills have indicated that, while landfills have performed relatively well, cracking and some downslope movement of interim cover soils and disruption of gas collection systems have occurred. The effect of the seismic motions on the integrity of covers incorporating compacted clay layers or geosynthetics has not been well documented. However, there is concern that compacted clay layers could be prone to cracking and that displacements may occur along interfaces between two geosynthetics and between a geosynthetic material and soil.

· There is considerable uncertainty about the dynamic properties of wastes, especially municipal solid waste, and about the appropriate shear strength values to use for geosynthetic materials under dynamic loading.

· Allowable deformations of covers under seismic conditions are based on practical considerations rather than rigorous analysis. For non-critical covers, seismic deformation of covers may be handled as a maintenance issue.

· Water balance models that are used to predict the performance of covers in terms of water percolation through the cover contain numerous simplifying assumptions and have been inadequately verified by field data. Since these models provide wide-ranging estimates of water infiltration under site-specific conditions, their main value may be to compare alternative designs using different cover configurations and materials.

Based on the field monitoring of test covers and cover failures, the following remarks are made regarding cover performance:

· Several examples of inadequately protected compacted clay barrier layers that degraded after a few years as a result of desiccation, root penetrations, or both are known within the industry. Although covering a compacted clay layer with a geomembrane provides greatly improved protection, one case suggested that thermally induced flow could eventually desiccate even a geomembrane-covered compacted clay barrier layer.

TABLE 6.2. Factors Affecting Final Cover Performance [after Daniel and Gross (1996)]

Layer	Factor
Surface Layer	· Erosion by water and/or air · Evapotranspiration · Native versus exotic vegetation · Appropriate armoring for side slopes at arid sites
Protection Layer	· Erosion by water · Slope failures due to pore pressure buildup · Animal burrows · Deep root penetration
Drainage Layer	· Excessive clogging · Insufficient flow rate capacity · Insufficient drainage layer outlets or capacity
Barrier Layer	· Cracking due to desiccation, deformations from waste settlement, or seismic motions (e.g., with clay) · Root penetration · Resistance to gas migration · Slope stability · Creep of all materials (clay, bentonite, geomembrane, asphalt) · Service life (geomembranes and asphalt)
Gas Collection Layer	· Adequate cover over waste
Foundation Layer	· Adequate strength · Proper grading

- There are few data on the performance of covers incorporating geomembranes and compacted clay layers and even fewer data for covers that include capillary barriers or employ surface vegetation to enhance evapotranspiration.
- The performance life of final covers has not been established. Although, the service life of some components of the system can be estimated, the functional life of the surface layer and barrier layer are not well documented and the long-term performance of constructed cover systems has not been adequately documented.
- A number of cases of cover failures have been documented; however, most of these failures could have been prevented through proper design and construction. It is felt that some failures may have resulted from preoccupation with regulatory compliance rather than engineering design considerations.

The primary factors adversely affecting cover performance for each of the basic layer components in a final cover are summarized in Table 6.2.

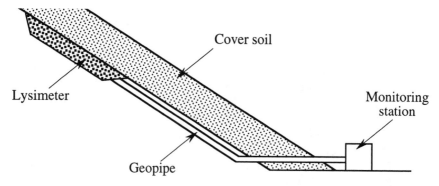

FIG. 6.6. *Lysimeter to Monitor Percolation through Cover Soil*

6.8 FIELD MONITORING

Final cover systems have generally been constructed with little or no instrumentation or monitoring. Because of the lack of monitoring, field performance data are generally limited to special test facilities. More information on the actual performance of final cover systems could be extremely beneficial in improving our regulations and designs of such systems.

One of the main objectives of final cover systems is to limit percolation of water into underlying waste. Monitoring of actual percolation rates is comparatively easy and inexpensive if lysimeters (Fig. 6.6) are used. A lysimeter is constructed with an impervious lining material (e.g., geomembrane) that is backfilled with granular soil or is covered with a geosynthetic. A geonet/geotextile composite drainage material may be preferred because it has much lower storage capacity than a thicker layer of sand or gravel. The lysimeter is drained by gravity to a monitoring point, where the flow is collected and periodically measured. Lysimeters can measure a meter or two in width, or can be much larger, measuring 10+ meters in width. Lysimeters provide an extremely powerful monitoring tool that measures directly the flux of liquid out the bottom of the final cover system.

Although total settlement and rate of settlement has been measured at many landfills, rarely has the differential settlement been measured or documented. The differential settlement, such as craters or sinkholes, is the most damaging type of settlement in cover systems. More information on the magnitude of the most severe differential settlement in landfill cover systems would be extremely valuable.

6.9 REFERENCES

Bonaparte, R., Othman, M. A., and Gross, B. A. (1977). "Preliminary Results of Survey of Field Liner System Performance," Proc. GRI-10

Conference on Field Performance of Geosynthetics, GII Publ., Philadelphia, PA, pp. 115–142.

Boschuk, Jr., J. (1991). "Landfill Covers An Engineering Perspective," Geotechnical Fabrics Report, Vol. 9, No. 4, IFAI, pp. 23–34.

Cheng, S. C. J., Corcoran, G. T., Miller, C. J., and Lee, J. Y. (1994). "The Use of a Spray Elastomer for Landfill Cover Liner Applications," Proc. 5th IGS Conference, Singapore, pp. 1037–1040.

Daniel, D. E., and Gross, B. A. (1996). Caps, Section 6, in *Assessment of Barrier Containment Technologies*, R. R. Rumer and J. K. Mitchell, Eds., National Technical Information Services (NTIS), PB96-180583, Springfield, VA, pp. 119–140.

Dwyer, S. F. (1995). "Alternative Landfill Cover Demonstration," Landfill Closures Environmental Protection and Land Recovery, Geotechnical Special Publication No. 53, R. J. Dunn and U. P. Singh, Eds., ASCE, pp. 19–34.

Gasper, A. J. (1990). "Stabilized Foam as Landfill Daily Cover," Proc. MSW Management: Solutions for the 90's, U.S. EPA, Washington, DC, pp. 1113–11121.

Gee, G. W., Freeman, H. D., Walters, Jr., W. H., Ligotkr, M. W., Campbell, M. D., Ward, A. L., Link, S. O., Smith, S. K., Gilmore, B. G., and Romine, R. A. (1994). "Hanford Prototype Surface Barrier Status Report: FY 1994," PNL-10275, Pacific Northwest Laboratory, Richland, Washington.

Gray, D. H., and Sotir, R. B. (1996). *Biotechnical and Soil Bioengineering Slope Stability*, J. Wiley and Sons, Inc., New York, NY, 310 pgs.

Horvath, J. S. (1995). "EPS Geofoam: New Products and Marketing Trends," GFR, Vol. 13, No. 6, IFAI, St. Paul, MN, pp. 22–26.

Horvath, J. S. (1994). Proc. Intl. Geotechnical Symposium on Polystyrene Foam in Below Grade Applications, Honolulu, Hawaii, Res. Rept. CE/GE-94-1, Manhattan College, Bronx, NY.

Khire, M. V. (1995). "Field Hydrology and Water Balance Modeling of Earthen Final Covers for Waste Containment," Environmental Geotechnics Report No. 95-5, University of Wisconsin-Madison, 166 pgs.

Koerner, R. M., Carson, D. A., Daniel, D. E., and Bonaparte, R. (1996). "Current Status of the Cincinnati GCL Test Plots," Proc. GRI-10 on Field Performance of Geosynthetics and Geosynthetic Systems, Geosynthetic Information Institute, Philadelphia, PA, pp. 153–182.

Koerner, R. M., and Daniel, D. E. (1994). "A Suggested Methodology for Assessing the Technical Equivalency of GCLs to CCLs," Proc. GRI-7 on Geosynthetic Liner Systems, IFAI Publ., St. Paul, MN, pp. 265–285.

Melchoir, S., Berger, K., Vielhaber, B., and Miehlich, G. (1994). "Multilayered Landfill Covers: Field Data on the Water Balance and Liner

Performance," *In-situ Remediation: Scientific Basis for Current and Future Technologies,* G. W. Gee and N. R. Wing, eds., Battelle Press, pp. 411–425.

Melchoir, S., and Miehlich, G. (1989). "Field Studies on the Hydrological Performance of Multilayered Landfill Caps," Proceedings of the Third International Conference on New Frontiers for Hazardous Waste Management, EPA/600/9-89/072, USEPA, pp. 100–107.

Miller, S. H. (1996). "Environmental Mine Design and Implications for Closure," Land and Water, September/October, pp. 32–34.

Moo-Young, H. K., and Zimmie, T. F. (1995). "Design of Landfill Covers Using Paper Mill Sludges," Proceedings of Research Transformed into Practice Implementation of NSF Research, J. Colville and A. M. Amde, eds., ASCE, NY, pp. 16–28.

Narejo, D. B., and Shettima, M. (1995). "Use of Recycled Automobile Tires to Design Landfill Components," Geosynthetics Intl., Vol. 2, No. 3, pp. 619–625.

Peggs, I. D. (1996). "Defect Identification, Leak Location and Leak Monitoring in Geomembrane Liners," Proc. Geosynthetics: Applications, Design and Construction, DeGroot, Den Hoedt, and Termaat, Eds., Balkema Publishers, Rotterdam, pp. 611–618.

Peterson, K. L., Link, S. O., and Gee, G. W. (1995). "Hanford Site Long-Term Surface Barrier Development Program: Fiscal Year 1994 Highlights," PNL-10605, Pacific Northwest Laboratory, Richland, Washington.

PIANC. (1996). Permanent International Association of Navigation Congresses, Report of Working Group No. 12, Brussels, Belgium, 36 pgs.

Pohland, F. G. (1996). "Landfill Bioreactors: Historical Perspectives, Fundamental Principles, and New Horizons in Design and Operations," EPA/600/R-95/146, September, pp. 9–24.

Primeaux, D. J. (1991). "100% Solids Aliphatic Spray Polyurea Elastomer Systems," *Polyurethanes World Congress 1991, SPI/SOPA,* Nice, France.

Rodël, A. (1996). "Geologger—A New Type of Monitoring System for the Total Area Monitoring of Seals on Landfill Sites," Proc. Geosynthetics: Applications, Design and Construction; DeGroot, Den Hoedt and Termaat, Eds., Balkema Publishers, Rotterdam, pp. 625–626.

Schultz, R. K., Ridky, R. W., and O'Donnell, E. (1995). "Control of Water Infiltration into Near Surface LLW Disposal Units, Progress Report of Field Experiments at a Humid Region Site, Beltsville, Maryland," U.S. Nuclear Regulatory Commission, NUREG/CR4918 Vol. 8, 20 pgs.

Shaw, P., and Carey, P. J. (1996). "Leachate Remediation Consideration for Design and Implementation," Conf. on Landfill, Design Construction and Operations, ESD, Detroit, March 18–20, 13 pgs.

Stormont, J. C. (1995). "The Performance of Two Capillary Barriers During Constant Infiltration," Landfill Closures Environmental Protection and Landfill Recovery, Geotechnical Special Publication No. 53, R. J. Dunn and U. P. Singh, Eds., ASCE, NY, pp. 77–92.

Theisen, M. S. (1992). "The Role of Geosynthetics in Erosion and Sediment Control: An Overview," Jour. Geotextiles and Geomembranes, Vol. 11, Nos. 4–6, pp. 199–214.

U.S. Environmental Protection Agency. (1996). *Landfill Bioreactor Design and Operation*, EPA/600/R-95/146, September, 230 pgs.

Virginia Erosion and Sediment Control Handbook, 3rd Edition, 1992, Virginia Department of Conservation and Recreation, Division of Solid and Water Conservation, Richmond, Virginia.

White, R. (1995). "EPS Geofoam: Unique Solutions for Forming Steep Landfill Embankments," Geosynthetics World, Vol. 5, No. 2, pg. 2.

Whitty, P. A., and Ballod, C. P. (1990). *Tire Chip Evaluation: Permeability of Leachability Assessments*, Final Report by Waste Management of North America, Inc. to Pennsylvania Dept. of Environmental Resources, Norristown, PA, February 28, 1990.

Wing, N. R., and Gee, G. W. (1994). "Quest for the Perfect Cap," *Civil Engineering*, ASCE, Oct., pp. 38–41.

OTHER CONSIDERATIONS AND SUMMARY

There are a number of considerations which are critical to the success of final cover systems that have not yet been addressed. These include quality control and quality assurance, lifetime assessment of the various materials, warrants for geosynthetic materials, and post-closure care/escrowed funds for long-term maintenance. Sections will be devoted to each topic in this chapter. A summary section will conclude the book and will include selected research needs from the perspective of the authors.

7.1 CURRENT CONCEPTS REGARDING QUALITY, e.g., ISO 9000

Quality in manufacturing and construction refers to an organization's desire to provide the best possible product in the context of their stated quality objectives. This is usually set forth in the organization's quality control (QC) manual, but is really much more than simply a document written by management within a company or organization. The essence of quality control is a top-to-bottom philosophy which is embodied from the company president through the ranks of personnel to the newest entry-level employee. When one refers to a specific structure for quality control, the program described by the International Organization for Standardization (ISO) often comes to mind.

The ISO 9000 series consists of five separate standards that establish the requirements for quality systems of companies and other organizations. Compliance with the requirements provides assurance that the processes and systems used to provide products and services do, in fact, consistently produce the level of quality that is specified and/or advertised. The standards are generic and specify the necessary elements of a quality system.

The ISO 9000 standards were developed during the 1980s. They evolved from previously established quality system standards. In particular, ISO 9000 was influenced by BSI-5750, developed by the British Standards Institution, and Mil-Q 9858A, used by the United States Department of Defense (Jenkins 1993).

The standards were given great credibility when they were adopted by the European Community (EC). The ISO 9000 series (or its European equivalent, the European Normalization Standard EN 29000 series) was seen by the EC as a means for guaranteeing cross-border quality between the countries of that community. European buyers and regulators regularly require the use of a third party registrar to ensure conformance to the appropriate ISO 9000 requirements. Third party registrars audit a company to ensure that its quality system, as documented and implemented, satisfies the requirements of the appropriate ISO 9000 standard.

There are a total of five standards. The first in the series, i.e., ISO 9000, is the road map. It also defines key terms, definitions, and related information. ISO 9001, 9002, and 9003 are quality standards specifying quality system models to be used in contractual situations. These three standards consist of a nested series as shown in Figure 7.1. ISO 9004 provides guidance for implementation.

- ISO 9001 is the most comprehensive. It is used to assure quality in design, development, production, installation, and service. It is typically used by organizations, such as automobile manufacturers, that design their own products and manufacture them accordingly.

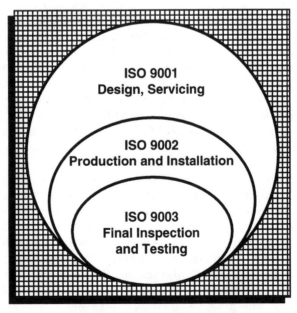

ISO 9001
Design, Servicing

ISO 9002
Production and Installation

ISO 9003
Final Inspection
and Testing

FIG. 7.1.   ISO 9000 Quality Standard [Jenkins (1993)]

- ISO 9002 is used when conformance to specified production and installation requirements needs to be assured—for example, when products are manufactured to specifications provided by outside agencies or contractors. Process industries, such as polymer and geosynthetic materials manufacturing, rely exclusively on ISO 9002.
- ISO 9003 is the least detailed standard. It requires only that conformance in final test and inspection be assured. It is typically used by such organizations as testing laboratories, calibration houses, and equipment distributors that inspect and test supplied products.
- ISO 9004 is used internally and not in contractual situations. It lists the essential elements that make up a complete quality system, including the responsibilities of management, marketing, procurement, corrective action, human-resource use, product safety, and use of statistical methods. Many of these elements are elaborations of concepts contained in the three previous standards.

Spizizen (1992) provides further detail on each of the above standards.

The ISO 9000 standards recognize 20 distinct elements in a quality system and list the requirements for each element in a separate section. The 20 elements are comprehensive in that they include every activity in an enterprise that affects quality, from design to contractual service. It should be noted that in the United Kingdom engineers, lawyers, doctors, and universities are seeking registration under the ISO 9000 process.

One of the most distinctive characteristics of ISO 9000 is the extensive documentation required. It requires complete and thorough documentation of all policies, objectives, activities, procedures, work rules, forms, and records that affect the quality system. It also requires that this documentation be maintained and updated.

Once a company or organization meets the objectives and criteria of the respective ISO 9000 category, that company or organization is said to be ISO-certified. However, the original designation, which was ISO-registered, seems more appropriate to the authors, since the products or services of the company or organization are not really certified, per se. What is achieved, and is most significant, is the realization that a quality system exists within the company or organization. This, in turn, strongly suggests that the company or organization can produce high quality materials or services.

7.1.1 Quality Control

As presented in the United States EPA Technical Guidance Document (EPA 1993), quality control refers to construction issues (hence CQC) when dealing with natural soil materials; and both manufacturing and construction issues when dealing with geosynthetics (hence MQC and CQC). In the context of this book on final cover systems, CQC rests within the earthwork contractor for compacted clay liners and sand/gravel drainage materials. Note that there is no MQC for natural soil materials since natural soils are not manufactured and are usually available on-site or transported to the site. CQC essentially begins the process for natural soil materials. With respect to geosynthetic materials, however, MQC rests with the product manufacturer and CQC rests with the installation contractor. More than one aspect of quality control can be performed by the same organization. For example,

- the earthwork contractor may also install the geosynthetics (which is not very common), and
- the geosynthetics manufacturer may also be the installer of the geosynthetics in the field (which is quite common).

It should be mentioned that in-house quality control plans and documents among contractors and installers of landfill cover systems are not common to such organizations. Conversely, quality control plans are commonplace for geosynthetic manufacturers, and many have ISO 9002 certification. This type of certification for geosynthetic manufacturers is particularly common in Europe and is a growing tendency in the United States. These efforts toward self-implementation of quality are very positive and are highly encouraged.

In spite of such growing activity in quality control, the tendency in federal and state environmental regulations, in both the United States and Germany, is to impose an additional level of monitoring on waste containment systems. This is called quality assurance.

7.1.2 Quality Assurance

As presented in the U.S. EPA Technical Guidance Document (EPA 1993), quality assurance pertains to the manufacturing of geosynthetic materials (hence MQA) as well as to the construction and installation of both natural soil materials and geosynthetics (hence CQA). In this context, MQA of geosynthetics rests with a separate group within the manufacturing organization (sometimes called "internal MQA") or by a separate consulting group (sometimes called "external MQA"). In either case, the MQA group should obviously have geosynthetic manufacturing experience. CQA of both natural soil materials and geosynthetics almost always rests with a separate consulting group. While the word "inspection" is usually associated with such MQA and CQA groups, it is more accurately characterized as "monitoring."

TABLE 7.1. Certification Levels for CQA and CQC Organizations Recommended by the U.S. EPA (EPA, 1993)

No. of Field Crews at Each Site	CQA Monitoring Organizations	CQC Installation Organizations
1–2	1—Level III	1—Level III
3–4	1—Level III	1—Level III
	1—Level I	
≥5	1—Level III	1—Level III
	1—Level II	1—Level I
	1—Level I	

Notes:
Level I = entry level person with nominal experience and a limited number of successful test results
Level II = 2-year experience level person with a moderate number of successful test results
Level III = 5-year experience level person with a large number of successful test results *or* a Professional Engineer with applicable geosynthetics experience

Due to the importance of the proper installation of geosynthetics, and their relative newness as a construction material, the selection of a consulting organization for CQA monitoring is critical. The U.S. EPA has recommended that a certain number of individuals be certified at least to the levels presented in Table 7.1. Note that the table also includes recommendations for CQC personnel, but requires a lesser number of individuals than for CQA. Tests for the above certification levels for geosynthetics monitoring and installation personnel are available through the National Institute for Certification of Engineering Technologies (NICET) in Alexandria, Virginia. A similar program for certification testing of monitoring and installation personnel of natural soil materials is being developed by the same organization. There is also an emerging certification testing program specifically for CQC personnel. It is being prepared by the International Association of Geosynthetic Installers (IAGI) in St. Paul, Minnesota.

7.2 LIFETIME

In addition to the myriad technical issues discussed in this book, perhaps the most contentious issue of all is the required lifetime for a landfill closure to perform its intended function. The issue is indeed technical, but goes far beyond physical quantification and often involves social, economic, political, and legal aspects, as well as technical. This section, however, will dwell only on technical issues, since they are essentially material-dependent. In this section, both natural soil materials and geosynthetics will be discussed.

7.2.1 Natural Soil Materials

Natural soil materials play an essential role in the long-term performance of final closures of landfills, abandoned dumps, and remediation projects. Natural soils are the essential part of the surface soil, protection soil, drainage soil, compacted clay barrier, gas venting layer (when soil), and foundation layer. As such, the use of soil represents as permanent an engineering material as one can possibly have, i.e., soils are indeed geologic materials insofar as their durability and lifetime is concerned. There are two aspects of natural soils, however, in which some discussion is necessary. They are the water within a compacted clay liner and the durability of some types of drainage soils.

7.2.1.1 Compacted Clay Liner Soils.   As was seen in the various final cover cross sections in Chapter 3, compacted clay liners form an essential barrier material in many cases. Regulations often prescribe their use. In discussing the lifetime of compacted clay liners, it must be recognized that their low hydraulic conductivity arises from a soil/water mixture which must be kept at, or near, complete saturation. In order to achieve low hydraulic conductivity, compacted clay is placed at water contents that are usually wet of optimum (EPA 1993). After compaction, the voids of the soil are usually 80 to 90% saturated with water.

Clearly, the soil component part of a compacted clay liner will last for geologic time. It is the water part which can create problems. In arid areas, or areas which are seasonally dry, the water in the voids of the soil can gradually be lost via evaporation or by migration of water to other soils or materials. As drying progresses, shrinkage occurs and reaches a limit at which cracking can occur. This cracking, caused by desiccation, occurs in block form, and gradually progresses deeper into the compacted clay liner until a pathway of water migration becomes available. The dry density of the soil has almost no effect on the vulnerability of the soil to desiccation cracking; water content is the master variable that dominates susceptibility to cracking (Daniel and Wu 1993). Highly plastic clays undergo large shrinkage when dried; clayey sand undergoes relatively little shrinkage. Shrinkage and cracking can occur as a result of changes of water content of only 2 to 5 percentage points. Moisture variations of this magnitude are inevitable in the top 1 to 2 m of soil at most sites. Over periods of decades or centuries, variation in moisture of 2 to 5 percentage points can occur to even greater depths. With the reintroduction of water, swelling occurs and the cracks close. However, the degree to which cracks swell shut is very sensitive to overburden stress (Boynton and Daniel 1985). At overburden stresses of less than 40 to 100 kPa, the cracks remain, even after the soil is soaked. The overburden stress on clay liners in landfill covers is usually ≤25 kPa. Thus, in final cover systems, the remnants of desiccation cracks are very likely to remain, causing the hydraulic conductivity to increase over its original value.

Freezing temperatures can also cause cracking in compacted clay liners (Othman et al. 1993). As with desiccation cracks, cracks induced by freeze-thaw cycles do not fully heal when the clay is soaked at low overburden stress. Thus, if the compacted clay liner is subjected to freeze-thaw, the liner will very likely not have its as-placed and intended low hydraulic conductivity. The one exception to this statement is compacted soil-bentonite admixtures, which do not appear to be vulnerable to damage from freeze-thaw action (Wong and Haug 1991).

The lifetime of a compacted clay liner (considering it to be a soil/water composite mixture) is clearly material- and site-specific. The factors that affect the service life of compacted clay liners (CCLs) are summarized in Table 7.2.

7.2.1.2 Sand/Gravel Drainage Soils.
In the consideration of natural soil drainage materials, the issue of lifetime of the soil particles is again a moot point, assuming that the soil is composed of stable minerals (e.g., quartz) rather than potentially soluble materials, such as limestone. What is an issue with some types of sand and gravel is the possibility of precipitates leaching from the soil particles and forming an encrustation at the particle-to-particle contacts. Over time, this encrustation can decrease the original hydraulic conductivity of the drainage material. Drainage soils

TABLE 7.2. Factors Affecting Service Life of Compacted Clay Liners (CCLs)

Factors Promoting a Long Service Life for CCLs:	Factors Likely to Lead to a Short Service Life for CCLs:
• Burial of CCL deep (>1 to 2 m) beneath surface of cap	• Placement of CCL close to ground surface (i.e., burial with <1 m of cover soil)
• Protection against desiccation provided by a geomembrane or other type of vapor barrier	• No geomembrane or other vapor barrier provided
• Use of clayey sand	• Use of highly plastic clay or other type soil likely to undergo large shrinkage when dried
• Placement and compaction of soil at relatively low water content	• Placement and compaction of soil at relatively high water content
• Climate with high rainfall year-round, drought periods of short duration	• Highly variable rainfall, with prolonged droughts occasionally occurring
• Cool climate that minimizes evapotranspiration	• Periods of year with warm temperature and very high evapotranspiration

composed of limestone, dolostone, dolomite, calcite, or other carbonates may create problems, if not from dissolution of the soil, then from precipitation of solids. A maximum carbonate content value should be specified with the amount set for site-specific conditions. A common maximum percentage of carbonates is 20%. This issue often results in the use of coarser drainage soil, i.e., larger particle sizes, than would be required for short-term drainage considerations.

Regarding a quantification of the durability of drainage soils, the U.S. Corps of Engineers (1996) uses the Los Angeles abrasion test (ASTM C131) and an alkaline immersion test for particle soundness (ASTM C88).

The *Los Angeles abrasion test* is a measure of aggregate degradation resulting from a combination of actions, including abrasion and attrition (friction). The test is conducted in a rotating steel drum containing a specified number of steel spheres, depending on the aggregate gradation. As the drum rotates, a shelf plate within the drum collects the aggregate and steel spheres allowing them to drop against the opposite side of the drum upon rotation, thus creating an impact-crushing effect. The aggregate and steel spheres roll within the drum under an abrading and grinding action until collected by the steel shelf for a subsequent revolution or cycle. Typically, 500 drum revolutions are specified. Loss in aggregate weight as a percentage of the original test sample weight is calculated and reported as the loss due to degradation.

Soundness testing measures resistance to weathering. This is accomplished by repeated immersion of an aggregate sample in saturated solutions of sodium or magnesium sulfate followed by oven drying to partially or completely dehydrate the salt precipitated in pore spaces within the aggregate. Upon rehydration in the salt solution, expansive forces occur which simulate the expansive forces exerted on the aggregate by freezing water. Testing is based on five immersion cycles in magnesium sulfate, as crystallization of magnesium sulfate results in greater stresses on aggregate than sodium sulfate. Following completion of the five immersion cycles, aggregate samples are washed and sieved to determine percentage loss due to soundness.

Maximum allowable percentage losses from abrasion (Los Angeles) and weathering (soundness) testing consistent with ASTM C33, Class 5S recommended by the Corps of Engineers are:

- Los Angeles: ≤50% loss
- Soundness: ≤18% loss

7.2.1.3 Sand Filtration Soils. Experience has shown that the most common cause of failure of drainage systems in landfill covers is a result of failure to provide an adequate filter. When drainage materials are placed adjacent to soils, an adequate filter should **always** be provided. Filter criteria were summarized in Chapter 2 for both natural soils and geotextiles. If the drainage material does not meet filter criteria for the adjacent soils, then one or more filters should be provided. This is usually the case. If an adequate filter is provided, the service life of the filter should not be an issue.

7.2.2 Geosynthetics

The aging process for geosynthetic materials is relatively unique, at least with respect to the aging of soils. The degradation of geosynthetic materials involves a gradual transition from a ductile material to a brittle one. As embrittlement occurs, the geosynthetic material does not disappear, but settlement, deformation, seismic vibration, etc., could cause a brittle cracking, signifying the end of the material's functional service life. Clearly, such failure mobilizing mechanisms can be envisioned in the final cover of a landfill, abandoned dump, or remediation project.

While the degradation mechanisms leading to this embrittlement are many, the most severe ones are eliminated by timely covering of the geosynthetics. For example, degradation by ultraviolet light and elevated temperature are eliminated in this manner. Furthermore, chemical attack in the cover of a landfill, abandoned dump, or remediation project is not likely since the location is above the waste. This leads to the primary mechanism of concern being *oxidation* of the polymer, causing embrittlement over a long time period.

Conceptually, the oxidation of geosynthetics can be considered in three distinct stages. These stages are designated as (a) depletion time of antioxidants, (b) induction time to the onset of polymer degradation, and (c) degradation of the polymer to decrease some property(s) to an arbitrary level, e.g., to 50% of its original value, see Figure 7.2.

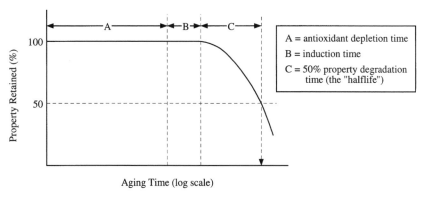

FIG. 7.2. Three Conceptual Stages in Chemical Aging of HDPE Geomembranes

7.2.2.1 *Depletion of Antioxidants.* The purpose of antioxidants in a geosynthetic formulation is to prevent degradation during processing and to prevent oxidation reactions taking place during the first stage of service life. However, there is only a limited amount of antioxidants in any formulation. Hence the lifetime for this stage is limited to the specific amount of antioxidant used. Once the antioxidants are completely depleted, oxygen will begin to attack the polymer, leading to the induction time stage and subsequently to the degradation of performance properties. The duration of the antioxidant depletion stage also depends on the type of selected antioxidants. Since many different antioxidants can be selected, depletion time can vary from formulation to formulation, subsequently affecting the lifetime of the geomembrane. Proper selection of antioxidants is known to contribute greatly to the overall lifetime of the geomembrane. For example, for a high density polyethylene (HDPE) geomembrane formulation with approximate 0.5% antioxidant package, Hsuan and Koerner (1996) have found that the times for antioxidant depletion are as follows:

> (a)   For water immersion with the water constantly in motion, as possibly occurs in a surface impoundment:
> Antioxidant Depletion Time = 41 to 44 years at 25°C in-situ temperature.
> (b)   For a simulated landfill environment under compressive stress with dry sand below and 300 mm of water above the geomembrane:
> Antioxidant Depletion Time = 126 to 128 years at 25°C in-situ temperature.

Note that these times are for HDPE, which is arguably the most stable of polymers being used in geosynthetics. In the United States, HDPE is the geomembrane of choice beneath waste. Above the waste a variety of polyethylene, polypropylene, and polyvinyl chloride liners are currently in use. HDPE is the only polymer allowed for use in waste containment liners **and covers** in Germany.

The above study has been ongoing for approximately five years and only recently has the data on antioxidant depletion time become available. Other geosynthetics are being evaluated in a like manner. Incubation in elevated temperature water baths and forced air ovens is ongoing for:

> (a)   high density polyethylene (HDPE) geogrids,
> (b)   polyester (PET) geogrids and geotextiles, and
> (c)   polypropylene (PP) geotextiles.

Whatever the polymer and its incubation method, the depletion of anti-oxidant leads to the induction stage.

7.2.2.2 Induction Time. In properly formulated geosynthetics, i.e., with resin, antioxidants, and carbon black, oxidation begins to occur after the depletion of the antioxidant. It occurs extremely slowly in a buried environment. The process can be visualized by tracking the oxygen absorption curve as diagrammed in Figure 7.3.

The initial position of the oxygen absorption curve (after depletion of the antioxidant) is called the *induction stage*. It is the time period in which there is no measurable change in the physical-mechanical properties of the geosynthetic material. The reason for this is explained via the following chemical oxidation reactions.

The first step of oxidation (after depletion of the antioxidants) is the formation of free radicals. The free radicals subsequently react with oxygen and start chain reactions. The reactions are described in Equations 7.1 to 7.6, after Grassie and Scott (1985).

Initiation stage:

$$RH \rightarrow R\bullet + H\bullet \text{ (under energy or catalyst residues)} \qquad (7.1)$$

$$R\bullet + O_2 \rightarrow ROO\bullet \qquad (7.2)$$

Propagation stage:

$$ROO\bullet + RH \rightarrow ROOH + R\bullet \qquad (7.3)$$

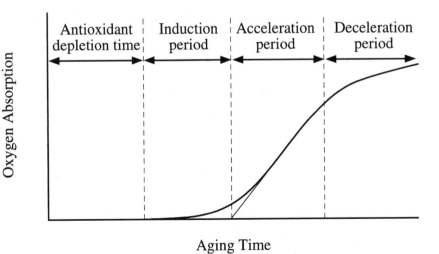

FIG. 7.3.   *Curves Illustrating Various Stage of Oxidation*

Acceleration stage:

$$ROOH \rightarrow RO\bullet + OH\bullet \text{ (under energy)} \qquad (7.4)$$

$$RO\bullet + RH \rightarrow ROH + R\bullet \qquad (7.5)$$

$$OH\bullet + RH \rightarrow H_2O + R\bullet \qquad (7.6)$$

In the above equations, RH represents the polyethylene polymer chains and the symbol"$\bullet$" represents free radicals. The free radicals are highly reactive in that they cause chain scission of the polymer backbone, which eventually results in embrittlement of the material.

In the induction stage, little hydroperoxide (ROOH) is present, and when formed, it does not decompose. Thus an acceleration stage of the oxidation cannot be achieved. As oxidation propagates slowly, additional ROOH molecules are formed. Once the concentration of ROOH reaches a critical level, decomposition of ROOH begins and accelerated chain reactions begin. This signifies the end of the induction period (Rapoport and Zaikov, 1986). This also indicates that the concentration of ROOH has a major effect on the duration of the induction period.

Viebke et. al. (1994) have studied the induction time of an unstabilized medium density polyethylene pipe. The pipes were internally pressure tested with water and externally by circulating air at temperatures ranging from 70 to 105°C. They found the activation energy of oxidation in the induction period to be 75 KJ/mol. Using their experimental values for the material evaluated, an induction time of 12 years was extrapolated at a typical in-service temperature of 25°C. This value is conservative with respect to a set of 20-year old HDPE water and milk bottles exhumed from a landfill, which showed no signs of degradation based on the lack of changes in the yield stress/yield strain/modulus values of the material. There was a decrease of approximately 30% in the break strength/break elongation values.

7.2.2.3 Polymer Degradation.
The end of the induction period signifies the onset of relatively rapid oxidation. This is the third, and final, stage in geosynthetic degradation. This oxidation is because the free radicals increase significantly due to the decomposition of ROOH, as indicated in Equations 7.4 to 7.6. One of the free radicals is an alkyl radical (R$\bullet$) which represents polymer chains that contain a free radical. In the early stage of acceleration, cross-linking occurs in these alkyl radicals due to oxygen deficiency. The reactions involved are expressed by Equations 7.7 and 7.8. The physical and mechanical properties of the material subsequently respond to such molecular changes. The most noticeable change is in the melt index, since it relates to the molecular weight of the polymer. In this stage, a lower melt index value is detected. By contrast, the mechanical

properties do not seem to be very sensitive to cross-linking. The tensile properties generally remain unchanged or are undetectable.

$$— CH_2 - \overset{\bullet}{C}R_1 - CH_2 —$$

RO• /   \ x2

$$— CH_2 - CR_1 - CH_2 —$$
$$\underset{|}{O}$$
$$— CH_2 - \overset{\bullet}{C}R_1 - CH_2 —$$
(Equation 7.7)

$$— CH_2 - CR_1 - CH_2 —$$
$$|$$
$$— CH_2 - CR_1 - CH_2 —$$
(Equation 7.8)

As oxidation proceeds further and abundant oxygen becomes available, the reactions of alkyl radicals change to chain scission. This causes a reduction in molecular weight, as shown in Equations 7.9 and 7.10. In this stage, the physical and mechanical properties of the material change according to the extent of the chain scission. The melt index value reverses from the previous low value to a value higher than the original starting value, signifying a decrease in molecular weight. As for tensile properties, break stress and break strain decrease. Tensile modulus and yield stress increase and yield strain decreases, although to a lesser extent. Eventually the geosynthetic material becomes brittle in that the tensile properties change significantly and the engineering performance is jeopardized, as described previously. This signifies the end of the so-called service life of the geomembrane.

$$— CH_2 - \overset{\bullet}{C}R_1 - CH_2 — \xrightarrow{O_2 \ \& \ RH} — CH_2 - CR_1 - CH_2 —$$
$$\underset{OOH}{|}$$

(7.9)

$$— CH_2 - CR_1 - CH_2 — \ + \ \overset{\bullet}{O}H$$
$$\underset{O\bullet}{|}$$

$$— CH_2 - CR_1 - CH_2 — \longrightarrow — CH_2 - CR_1 - O + \bullet CH_2 —$$
$$\underset{O\bullet}{|}$$

(7.10)

Although quite arbitrary, the end of service life of a geosynthetic material is often selected as a 50% reduction in an important design property. This is commonly referred to as the half-lifetime, or simply the "halflife." The specific property could be yield stress, yield strain, modulus of HDPE or the comparable break properties of geosynthetic resins which do not show a pronounced yield point. It should be noted that even at halflife, the geosynthetic still exists and can function, albeit at a decreased performance level. Koerner et al. (1991) and Hsuan et al. (1993) have used literature values for activation energy and found this stage of geosynthetic lifetime to possibly be a few hundred years.

7.2.2.4 Anticipated Lifetime. As just presented, the lifetime of a properly formulated geosynthetic, e.g., a HDPE geomembrane, will be equal to the depletion time of antioxidants, plus induction time, plus the time to reach a 50% reduction in a specific engineering property. Graphically this was shown in Figure 7.1 as the sum of "A," "B," and "C." To the best of our knowledge at this time, this value is as follows:

Stage A: Antioxidant Depletion Time = 50 to 150 years
Stage B: Induction Time = 10 to 30 years
Stage C: Time to Halflife = unknown, but likely to be a few hundred years

Thus the lifetime for a geosynthetic, such as a properly formulated HDPE geomembrane, will be at least a few centuries, and possibly 1000 years. The lifetime for other geosynthetics, particularly those with greater specific surface area than geomembranes, e.g., geotextiles, is less certain than the above estimate. While the lifetime for other geosynthetics could be envisioned as lower, this is not known to be the case. Work is ongoing in this regard.

7.3 WARRANTS

Of all of the different types of materials used in the construction of final covers for landfills, dumps, and remediation projects, only geomembranes have warrants associated with their use and performance. This peculiar situation arises from the use of similar polymeric materials on the roofs of buildings, i.e., for roofing membranes. A typical warrant for a flat roof membrane in the United States is 20 years. With such a warrant, the manufacturer guarantees the material for such a time period. If it does not perform properly, i.e., if the material degrades in a shorter period than the warrant states, the cost of the material is pro-rated and a refund (or discount on a new roofing membrane) is offered. By pro-rated, it is meant that if 15-years of satisfactory performance is obtained on a 20-year warrant, only 25% of the original price of the material is refunded or discounted on the new product. This is a material warrant only and installation costs are not involved. Due to the similarity of roofing membranes and geomembranes, the warrant concept has carried over into the waste containment industry. This is somewhat natural since some of the original manufacturers of geomembranes also sold product to the roofing industry, e.g., geomembranes like CSPE-R, CPE-R, and EPDM-R are also used for flat roof construction. Of these materials, only scrim reinforced chlorosulphonated polyethylene (CSPE-R) continues to be used in waste containment applications, and rarely for final covers.

In the authors' opinion, this carry-over of warrants from the roofing industry to the geomembrane industry is a mistake. The reason is simple. Roofing membranes are exposed; while geomembranes (in final cover applications) are not exposed. With exposed roofing membranes, the polymer must resist many types of degradation mechanisms, e.g.,

- ultraviolet light,
- elevated temperature,
- moving water,
- ice and snow (in northern climates),
- thermally induced cyclic stresses, and
- directly applied live (often impact) type loads

to an extent simply not present in the soil-covered conditions of the final cover at an engineered landfill, abandoned dump, or remediation project. In such a buried condition, it has been shown in Section 7.2.2.4, that the lifetime of the geomembrane will be hundreds of years. Thus a 20-year warrant is simply not an issue and is actually foolish in light of the anticipated lifetime of the geomembrane.

Furthermore, the offering of geomembrane warrants represents a non-level playing field for the various geomembrane manufacturers. Those manufacturers which are uninsured, or self-insured, will offer such warrant with no concern or without trepidation. Those manufacturers which are independently insured will have to purchase insurance and actually sequester funds for the warranted time period. Such action freezes funds from other uses and represents an uneven financial burden.

A different type of warrant, however, may be considered and, paradoxically, is rarely requested. This is not a material warrant as just described; rather, it is an installation warrant. The concern over geomembranes, *and all geosynthetic materials,* is not as much with their manufacture (recall the discussion on quality control and quality assurance of Section 7.1), as it is with their installation. Holes, tears, rips, punctures, etc., either defeat the use of geosynthetics or greatly limit the performance of the system. As a result, an owner or operator (private or public) of a facility can request an installation warrant on some, or all, of the geosynthetics. Such installation warrants are available. Generally, they are for a 1-year period after installation and are offered by the installation contractors. Note that this may be the manufacturer, but in many cases, it is an independent contractor or installation company.

Neither material nor installation warrants are available on natural soil materials, e.g., on compacted clay liners or granular soil drainage layers.

7.4 POST-CLOSURE ISSUES

Upon completion of construction of the final cover of an engineered landfill, abandoned dump, or remediation project, a number of post-closure issues arise. They are care/maintenance, financial responsibility, and guarantees for unforeseen events. These issues are very contentious and vary from country to country, state to state, and waste to waste.

Regarding engineered landfills, the United States regulations are quite detailed on MSW, but curiously silent on hazardous wastes. Regarding MSW, 40 CFR 265.147 states that financial responsibility for accidental and nonaccidental occurrences must be demonstrated by the landfill owner. The regulations are quite detailed, even to the required types of guarantees, letters of credit, surety bonds, and/or trust funds for liability coverage. The usual post-closure care period is 30 years after completion of the final cover. These federal regulations can be superseded by state regulations. For example, California's Title 14, Article 3.4 is rigorous, even to the manner of calculations of post-closure maintenance costs, including the projected costs for:

- final cover;
- final grading;
- drainage system;
- gas monitoring and control system;
- leachate control system;
- ground water monitoring system;
- security (e.g., fences, gates, and signs);
- vector and fire control; and
- litter control.

Furthermore, in Article 3.5, financial responsibility is described along with the methods of establishing a trust fund, enterprise fund, and guarantee.

To the authors, this amount of prescriptive detail is excessive. As with any other engineered system, of which the final cover of an engineered landfill, abandoned dump, or remediation project is a subset, many standard controls are in place for long-term concerns and financial recourse.

In the event of an unintended or intentional occurrence, disputes can be mutually settled between the parties involved. More contentious issues can go to mediation or can become the subject of litigation. By taking this approach, a less-prescriptive and more performance-oriented methodology can be created.

Regarding abandoned dumps and remediation projects, the situation is quite different since the owner and/or disposer of the waste usually cannot be identified. For CERCLA remediation projects in the United States, the final cover contractor is responsible for the first year after completion of the final cover. For the next 29 years, the state is responsible for maintenance. However, after 5 years, the state and EPA conduct a site visit. If problems are found and if they resulted from design or construction deficiencies, the state may solicit funds from the EPA for repairs. This appears to the authors to be a reasonable approach, and if the results are not satisfactory, the parties involved always have legal recourse.

7.5 SUMMARY

In closing this book on final covers of engineered landfills, abandoned dumps, and reclamation projects, we wish to re-emphasize the philosophy that we have taken throughout, as well as the presentation of some research needs to further the technology of final closure systems.

7.5.1 Final Comments

This book has attempted to provide a *technically-based* and *unified* approach toward the design of final covers for engineered landfills, abandoned dumps, and remediation projects. These two points require further emphasis.

In the first summarizing point, the book is *technically-based*. Detail, as complete as possible, has been provided. For example, presented herein was considerable detail on:

- composition of the various layers in Chapter 2,
- example cross sections in Chapter 3,
- water balance analyses in Chapter 4,
- slope stability considerations in Chapter 5, and
- emerging concepts and materials in Chapter 6.

This technically-based aspect of the book was tempered, however, with regulatory issues in Chapter 1 and other non-technical considerations in Chapter 7. Regarding the regulatory issues in Chapter 1, final cover requirements in the United States and Germany were presented, since these two countries arguably have the most carefully considered requirements for waste containment facilities. Regarding the non-technical considerations in

Chapter 7, construction issues and lifetime predictions were both seen to be significant with regard to the long-term performance of final covers. Related financial issues of post-closure care and maintenance were also addressed in Chapter 7.

In the second summarizing point, the book presents a *unified* approach toward the design of final covers. The actual waste materials underlying the final cover include the following:

- hazardous and nonhazardous landfills,
- landfills containing a host of other materials, e.g., incinerator ash, low level radioactive wastes, construction debris, etc.
- abandoned dumps, and
- remediation projects requiring a final cover.

The nature of the waste and its site-specific potential for groundwater and/ or air pollution is all-important in dictating the type of design for a final cover. While regulations attempt to consider this issue, e.g., Subtitle C and D categories in the United States and Categories I, II, and III in Germany, it is a most difficult task. The reasons are that categorization is inherently difficult and site-specific conditions vary considerably.

In this regard the authors generally favor performance-based regulations over prescriptive-based regulations. This, of course, places great reliance on the design engineer and the regulatory group that issues the required permit. (The latter is usually a state agency rather than a federal one.) This is the case in both the United States and Germany. The tacit assumption when favoring a performance-based regulation is that the engineering design group and state regulatory personnel are sufficiently versed to accomplish the task at hand. It is sincerely hoped that this book provides a step in the direction of providing the requisite knowledge.

7.5.2 Research Needs

To assume that all of the requisite knowledge in the design of final closures of engineered landfills, abandoned dumps, and remediation projects is currently available is clearly presumptuous. Thus, as a closure to the book, we present a list of research needs which are felt to be important in raising the level of knowledge higher than it is currently. Additional detail is available in Daniel and Gross (1996).

- Data are available that demonstrate that the performance of compacted clay barrier layers in covers will deteriorate over time. Even so, compacted clay barrier layers are still being used, primarily because they are specified in regulations. There is a perception that it may be difficult to obtain regulatory approval to use alternative barrier materials. This is particularly the case for GCLs in the cover and, to a lesser extent, geomembranes. This situation could be improved if guidelines were available for demonstrating the equivalency of performance among the different options for cover components.

- Few data are available concerning the hydraulic performance of traditional covers with resistive barriers. There are even fewer performance data for covers containing capillary barriers. More data need to be collected to assess overall cover performance. Data are especially needed to bring about regulatory and community acceptance of alternative cover configurations. While some of these data can be collected from currently instrumented sites, other additional sites will probably need to be monitored.
- The expected performance lifetime of final covers is uncertain. Studies are underway to assess the service life of some individual cover components. However, the long-term service life of these components in a constructed system has not been adequately studied. Note that this involves all of the barrier components of the cover system, i.e., compacted clay liners, geosynthetic clay liners, geomembranes, and other ancillary geosynthetics.
- There have been a number of documented cases of cover failures; however, most of these failures could have been avoided through proper design and/or construction. There is a need for more education about the specific idiosyncrasies of cover design and for independent peer review of completed designs prior to construction. Compliance with regulations is not a sufficient check on a completed final cover design.
- More field observations on the effect of seismic motions on the integrity of covers incorporating compacted clay layers, geosynthetic clay liners, and/or geosynthetic materials need to be made.
- The shear strength at interfaces between materials (particularly geosynthetics) is known to be an important factor affecting the physical stability of covers on slopes. The shear strength is also affected by freezing-thawing cycles, heating-cooling cycles, and creep. Standard procedures for evaluating interface shear strength need to be developed.
- More information is needed on the dynamic shear strength values to be used for geosynthetic layer materials.
- More information is needed on the static and dynamic properties of wastes so as to determine total and differential settlements. The dearth of differential settlement data is of particular concern, since it dominates the design of the barrier materials.

- Other alternative barrier layer materials, such as asphalt and paper mill waste, may be possible for future use. However, more information needs to be collected on the long-term performance of these materials.
- Available computer models for simulating the hydrologic and hydraulic performance of covers need to be verified by comparison with field data, and modified as necessary.

7.6 REFERENCES

Boynton, S. S., and Daniel, D. E. (1985). "Hydraulic Conductivity Tests on Compacted Clay," Jour. of Geotechnical Engineering, Vol. 111, No. 4, pp. 465–478.

Daniel, D. E., and Gross, B. A. (1996). "Caps," Section 6 in Workshop Proceedings on Containment of Solid Waste, R. Rumer, Ed., NTIS Publ. No. PB96-180583, pp. 119–140.

Daniel, D. E., and Wu, Y. K. (1993). "Compacted Clay Liners and Covers for Arid Sites," Jour. of Geotechnical Engineering, Vol. 119, No. 2, pp. 223–237.

Grassie, N., and Scott, G. (1985). Polymer Degradation and Stabilization, Cambridge University Press, New York, U.S.A.

Hsuan, Y. G., and Koerner, R. M. (1996). "Long-Term Durability of HDPE Geomembranes: Part I—Depletion of Antioxidants," GRI Report #16, 37 pgs., Philadelphia, PA, December 11, 1995.

Hsuan, Y. G., Koerner, R. M., and Lord, Jr., A. E. (1993). "A Review of the Degradation of Geosynthetic Reinforcing Materials and Various Polymer Stabilization Methods," Geosynthetic Soil Reinforcement Testing Procedures, ASTM STP 1190, S. C. Jonathan Cheng, Ed., American Society for Testing and Materials, Philadelphia, PA, 1993, pp. 228–244.

Jenkins, M. (1993). "A Look at ISO 9000," ASTM Standardization News, July, pp. 50–52.

Koerner, R. M., Lord, Jr., A. E., and Halse-Hsuan, Y. (1991). "Degradation of Polymeric Materials and Products," Proc. Intl. Symp. on Research Development for Improving Solid Waste Management, U.S. EPA, Cincinnati, OH, pp. 1–11.

Othman, M. A., Benson, C. H., Chamberlain, E. J., and Zimmier, T. T. (1994). "Laboratory Testing to Evaluate Changes in Hydraulic Conductivity of Compacted Clays Caused by Freeze-Thaw: State-of-the-Art," Hydraulic Conductivity and Waste Containment Transport in Soil, ASTM STP 1142, American Society for Testing and Materials, Philadelphia, PA, pp. 227–251.

Rapoport, N. Ya., and Aaikov, G. E. (1986). "Kinetics and Mechanism of the Oxidation of Stressed Polymer," Developments in Polymer

Stabilization—4, Chapter 6, edited by Scott, G., Published by Applied Science Publishers Ltd., London, pp. 207–258.

Spizizen, G. (1992). "The ISO 9000 Standards: Creating a Level Playing Field for International Quality," National Productivity Review, Summer, 1992, pp. 331–346.

U. S. Corps of Engineers (1996). Draft Specifications for Rocky Mountain Arsenal Landfill, Denver, Colorado, November 18, 1996.

U.S. EPA (1993). "Quality Assurance and Quality Control for Waste Containment Facilities," Technical Guidance Document, EPA/600/R-93/182, September, 1993, 305 pgs.

Viebke, J., Elble, E., and Gedde, U. W. (1994). "Degradation of Unstabilized Medium Density Polyethylene Pipes in Hot Water Applications," Polymer Engineering and Science, Vol. 34, No. 17, pp. 1354–1361.

Wong, L. C., and Haug, M. D. (1991). "Cyclical Closed System Freeze-Thaw Permeability Testing of Soil Liner and Cover Materials," Canadian Geotechnical Journal, Vol. 28, pp. 784–793.

INDEX